Lecture Notes in Computer Science   5321

Commenced Publication in 1973
Founding and Former Series Editors:
Gerhard Goos, Juris Hartmanis, and Jan van Leeuwen

Nick Bassiliades   Guido Governatori
Adrian Paschke (Eds.)

Rule Representation, Interchange and Reasoning on the Web

International Symposium, RuleML 2008
Orlando, FL, USA, October 30-31, 2008
Proceedings

 Springer

Volume Editors

Nick Bassiliades
Aristotle University of Thessaloniki
Department of Informatics
Thessaloniki, Greece
E-mail: nbassili@csd.auth.gr

Guido Governatori
National ICT Australia
Queensland Research Laboratory
St Lucia, Queensland, Australia
E-mail: guido.governatori@nicta.com.au

Adrian Paschke
Free University Berlin
Chair for Corporate Semantic Web
Berlin, Germany
E-mail: paschke@inf.fu-berlin.de

Library of Congress Control Number: Applied for

CR Subject Classification (1998): D.3.1, F.3.2, H.5.3

LNCS Sublibrary: SL 2 – Programming and Software Engineering

ISSN   0302-9743
ISBN-10   3-540-88807-1 Springer Berlin Heidelberg New York
ISBN-13   978-3-540-88807-9 Springer Berlin Heidelberg New York

Springer is a part of Springer Science+Business Media

springer.com

© Springer-Verlag Berlin Heidelberg 2008

Typesetting: Camera-ready by author, data conversion by Scientific Publishing Services, Chennai, India
Printed on acid-free paper   SPIN: 12553501   06/3180   5 4 3 2 1 0

Preface

The 2008 International Symposium on Rule Interchange and Applications (RuleML 2008), collocated in Orlando, Florida, with the 11th International Business Rules Forum, was the premier place to meet and to exchange ideas from all fields of rules technologies. The aim of RuleML 2008 was both to present new and interesting research results and to show successfully deployed rule-based applications. This annual symposium is the flagship event of the Rule Markup and Modeling Initiative (RuleML).

The RuleML Initiative (www.ruleml.org) is a non-profit umbrella organization of several technical groups organized by representatives from academia, industry and government working on rule technologies and applications. Its aim is to promote the study, research and application of rules in heterogeneous distributed environments such as the Web. RuleML maintains effective links with other major international societies and acts as intermediary between various 'specialized' rule vendors, applications, industrial and academic research groups, as well as standardization efforts from, for example, W3C, OMG, and OASIS.

After a series of successful international RuleML workshops and then conferences, the RuleML Symposium, held since 2007, constitutes a new kind of event where the web rules and logic community joins the established, practically oriented business rules forum community (www.businessrulesforum.com) to help to cross-fertilize between web and business logic technology. The symposium supports the idea that there is a successful path from high-quality research results to applied applications. It brings together rule system providers, representatives of, and participants in, rule standardization efforts and open source rule communities, practitioners and technical experts, developers, users, and researchers, to exchange new ideas, practical developments and experiences on issues pertinent to the interchange and application of rules.

The technical program of RuleML 2008 showed a carefully selected presentation of current rule research and development in ten full papers, ten short papers, two demo papers, and four keynote talks (abstracts of three of them included) detailed in this book. Accepted papers covered several aspects of rules, such as rule engineering, rule representation languages, natural-language and graphical rule representation and processing, reasoning engines, rule-based methodologies in distributed and heterogeneous environments, rule-based applications for policies, electronic contracts and security. The papers were selected from 35 submissions received from 17 countries. RuleML 2008, as its predecessors, offered a high quality technical and applications program, which was the result of the joint effort of the members of the RuleML 2008 program committee.

The real success of rule technology will be measured by the applications that use the technology rather than the technology itself. To place emphasis on the practical use of rule technologies in distributed Web-based environments, the RuleML 2008 Challenge was a major International showcase of beneficial solutions for industry and commerce based on Web rules technologies. The applications covered a wide range of areas from industrial systems / rule engines and rule technologies, interoperation, and interchange. They illustrated not only the range of technologies being used in applications, but also the wide range of areas in which rules can produce real benefits. The challenge offered participants the opportunity to demonstrate their commercial and open source tools, use cases, and applications. It was the ideal forum for those wanting to understand how rules technology can produce benefits, both technically and commercially.

The RuleML 2008 organizers wish to thank the excellent program committee for their hard work in reviewing the submitted papers. Their criticism and very useful comments and suggestions were instrumental to achieving a high-quality publication. We also thank the authors for submitting good papers, responding to the reviewers' comments, and abiding by our production schedule. We further wish to thank the keynote speakers for their interesting talks. We are very grateful to the Business Rules Forum organizers for enabling this fruitful collocation of the 11th Business Rules Forum and RuleML 2008. Especially, we thank Gladys Lam and Valentina Tang for their support.

The RuleML 2008 Symposium was financially supported by industrial companies and research institutes and was techincally supported by several professional societies. We wish to thank our sponsors, whose financial support helped us to organize this event, and whose technical support allowed us to attract many high-quality submissions.

August 2008   Nick Bassiliades
                                                           Guido Governatori
                                                             Adrian Paschke

Symposium Organization

Organizing Committee

General Chair

Adrian Paschke · · · · · · Free University Berlin, Germany

Program Chairs

Nick Bassiliades · · · · · · Aristotle University of Thessaloniki, Greece
Guido Governatori · · · · · NICTA, Australia

Challenge Chairs

Costin Badica · · · · · · · University of Craiova, Romania
Yuh-Jong Hu · · · · · · · · National Chengchi University, Taiwan

Panel Chairs

John Hall · · · · · · · · · Model Systems, UK
Axel Polleres · · · · · · · DERI Galway, Ireland

Liaison Chairs

Mark Proctor · · · · · · · · JBoss Rules, UK
Rainer von Ammon · · · · · CITT GmbH, Germany
Jan Vanthienen · · · · · · · Katholieke Universiteit Leuven, Belgium

Publicity Chairs

Matthias Nickles · · · · · · University of Bath, UK
Tracy Bost · · · · · · · · · Valocity, USA

Web Chair

Suzanne Embury · · · · · · University of Manchester, UK

Program Committee

Asaf Adi	Ioannis Hatzilygeroudis	Antonio Rotolo
Grigoris Antoniou	Martin Hepp	Norman Sadeh
Sidney Bailin	Elisa Kendall	Christian de Sainte Marie
Matteo Baldoni	Yiannis Kompatsiaris	Marco Seiriö
Cristina Baroglio	Manolis Koubarakis	Timos Sellis
Claudio Bartolini	Alex Kozlenkov	Michael Sintek
Tim Bass	Holger Lausen	Silvie Spreeuwenberg
Bernhard Bauer	John Lee	Giorgos Stamou
Mikael Berndtsson	Mark Linehan	Giorgos Stoilos
Leopoldo Bertossi	Heiko Ludwig	Terrance Swift
Pedro Bizarro	Mirko Maleković	Kuldar Taveter
Peter Bollen	Christopher J. Matheus	James Taylor
Christian Brelage	Craig McKenzie	Vagan Terziyan
Donald Chapin	Jing Mei	Paul Vincent
Shyi-Ming Chen	Zoran Milosevic	George Vouros
Jorge Cuellar	Jang Minsu	Kewen Wang
Mike Dean	Leora Morgenstern	Mehmet Emre Yegen
Stan Devitt	Gero Muehl	
Jens Dietrich	Jörg Müller	
Jürgen Dix	Chieko Nakabasami	
Schahram Dustdar	Ilkka Niemelä	
Andreas Eberhart	Bart Orriens	
Opher Etzion	Jeff Pan	
Dieter Fensel	Paula-Lavinia Patranjan	
Dragan Gasevic	Jon Pellant	
Adrian Giurca	Jeff Pollock	
Stijn Goedertier	Alun Preece	
Robert Golan	Maher Rahmouni	
Christine Golbreich	Girish Ranganathan	
Tom Gordon	Dave Reynolds	
Marek Hatala	Graham Rong	

Additional Reviewers

Stamatia Dasiopoulou	Adrian Mocan
Verena Kantere	Yuting Zhao
Mick Kerrigan	

Sponsors

Gold Sponsor

Silver Sponsor

Bronze Sponsors

Cooperation Partners

IEEE SMCS TC on Intelligent Internet Systems
IEEE SMCS TC on Distributed Intelligent Systems

Media Partners

Table of Contents

Keynote Talks (Abstracts)

Rule Engineering

Rule-Based Methodologies and Applications in Policies, Electronic Contracts and Security

Rule Representation Languages and Reasoning Engines

Rule-Based Methodologies and Applications in Distributed and Heterogeneous Environments

Natural-Language and Graphical Rule Representation and Processing

RuleML-2008 Challenge

Rule Interchange Format: The Framework

(Extended Abstract)

Michael Kifer

State University of New York at Stony Brook, USA

The *Rule Interchange Format* (RIF) activity within the World Wide Web Consortium (W3C) aims to develop a standard for exchanging rules among disparate systems, especially on the Semantic Web. The need for rule-based information processing on the Web has been felt ever since RDF was introduced in the late 90's. As ontology development picked up pace this decade and as the limitations of OWL became apparent, rules were firmly put back on the agenda. RIF is therefore a major opportunity for the introduction of rule based technologies into the main stream of knowledge representation and information processing on the Web.

Despite its humble name, RIF is not just a format and is not primarily about syntax. It is an extensible framework for rule-based languages, called RIF *dialects*, which includes precise and formal specification of the syntax, semantics, and XML serialization. Extensibility here means that new dialects can be added, if sufficient interest exists, and the languages are supposed to share much of the syntactic and semantic apparatus.

Because of the emphasis on rigor and semantics, the term "format" in the name of RIF might seem a misnomer. However, making a cute acronym is not the only reason for the choice of this term. The idea behind rule exchange through RIF is that the different systems will be able to map their languages (or substantial parts thereof) to and from the appropriate RIF dialects in *semantics-preserving* ways and thus rule sets and data could be communicated by one system to another provided that the systems can find a suitable dialect, which they both support. The intermediate RIF language is supposed to be in the XML format, whence the term "format" in the RIF name.

The RIF Working Group has plans to develop two kinds of dialects: logic-based dialects and dialects for rules with actions. The logic-based dialects include languages based on first-order logic and also a variety of logic programming dialects based on the different non-first-order semantics such as the well-founded and stable semantics [6,12]. The rules-with-actions dialects will be designed for production rule systems, such as Jess and Drools [7,8], and for reactive rules such as those represented by XChange [3], FLORA-2 [9], and Prova [11]. At the time of this writing, only the *Basic Logic Dialect*, RIF-BLD (which belongs to the first category), has been substantially completed and is in the "Last Call" status in the W3C standardization process [1]. In the second category, a *Production Rule Dialect*, RIF-PRD, is under active development [5].

These plans are very ambitious, and in the beginning it was not at all obvious how the different dialects could be made to substantially share syntactic and, especially, semantic machinery. Even within the logic-based category the dialects

N. Bassiliades, G. Governatori, and A. Paschke (Eds.): RuleML 2008, LNCS 5321, pp. 1–2, 2008.
© Springer-Verlag Berlin Heidelberg 2008

are expected to have vastly different semantics: the first-order semantics warrants inferences that are different from those warranted by the logic programming semantics, and the various logic programming semantics do not agree in all cases. This is where the RIF extensibility framework comes in. At present, only the *Framework for Logic Dialects*, RIF-FLD, has been worked out to sufficient degree of detail [2], and this is the main subject of this paper.

This paper is an introduction to RIF Framework for Logic Dialects, an extensibility framework that ensures that the current and future dialects of the Rule Interchange Format share common syntactic, semantic, and XML markup apparatus. RIF-FLD is still work in progress: some details may change and additions to the framework should be expected.

Apart from RIF-BLD and the dialect under development for production rule systems, other dialects are being planned. These include the logic programming dialects that support well-founded and stable-model negation, a dialect that supports higher-order extensions as in HiLog [4], and a dialect that extends RIF-BLD with full F-logic [10] support (BLD accommodates only a very small part of F-logic).

The development of RIF is an open process and feedback from experts and users is welcome. All the documents of the working group, meeting agendas, and working lists are publicly available at the group's Web site http://www.w3.org/. 2005/rules/wiki/RIF_Working_Group. The working version of the RIF framework can be found at http://www.w3.org/ 2005/rules/wiki/FLD.

References

1. Boley, H., Kifer, M.: RIF Basic logic dialect. W3C Working Draft (July 2008), http://www.w3.org/TR/rif-fld/
2. Boley, H., Kifer, M.: RIF Framework for logic dialects. W3C Working Draft (July 2008), http://www.w3.org/TR/rif-fld/
3. Bry, F., Eckert, M., Patranjan, P.-L.: Reactivity on the web: Paradigms and applications of the language xchange. Journal of Web Engineering 5(1), 3–24 (2006)
4. Chen, W., Kifer, M., Warren, D.S.: HiLog: A foundation for higher-order logic programming. Journal of Logic Programming 15(3), 187–230 (1993)
5. de Sainte Marie, C., Paschke, A.: RIF Production rule dialect. W3C Working Draft (July 2008), http://www.w3.org/TR/rif-prd/
6. Gelfond, M., Lifschitz, V.: The stable model semantics for logic programming. In: Logic Programming: Proceedings of the Fifth Conference and Symposium, pp. 1070–1080 (1988)
7. Drools. Web site, http://labs.jboss.com/drools/
8. Jess, the rule language for the java platform. Web site, http://herzberg.ca.sandia.gov/jess/
9. Kifer, M.: FLORA-2: An object-oriented knowledge base language. The FLORA-2 Web Site, http://flora.sourceforge.net
10. Kifer, M., Lausen, G., Wu, J.: Logical foundations of object-oriented and frame-based languages. Journal of ACM 42, 741–843 (1995)
11. Kozlenkov, A.: PROVA: A Language for Rule-based Java Scripting, Data and Computation Integration, and Agent Programming (May 2005)
12. Van Gelder, A., Ross, K.A., Schlipf, J.S.: The well-founded semantics for general logic programs. Journal of ACM 38(3), 620–650 (1991)

The Power of Events: An Introduction to Complex Event Processing in Distributed Enterprise Systems

David Luckham

Department of Electrical Engineering, Stanford University, USA
luckham@stanford.edu

Abstract. Complex Event Processing (CEP) is a defined set of tools and techniques for analyzing and controlling the complex series of interrelated events that drive modern distributed information systems. This emerging technology helps IS and IT professionals understand what is happening within the system, quickly identify and solve problems, and more effectively utilize events for enhanced operation, performance, and security. CEP can be applied to a broad spectrum of information system challenges, including business process automation, schedule and control processes, network monitoring and performance prediction, and intrusion detection.

This talk is about the rise of CEP as we know it today, its historical roots and its current position in commercial markets. Some possible long-term future roles of CEP in the Information Society are discussed along with the need to develop rule-based event hierarchies on a commercial basis to make those applications possible. The talk gives empahsis to the point that "Rules are everywhere" and that mathematical formalisms cannot express all the forms that are in use in various event processing systems.

N. Bassiliades, G. Governatori, and A. Paschke (Eds.): RuleML 2008, LNCS 5321, p. 3, 2008.
© Springer-Verlag Berlin Heidelberg 2008

Event and Process Semantics Will Rule

Paul Haley

Chairman, Haley Systems, Inc.
President, Automata, Inc.
paul@haleyai.com

Abstract. The convergence of business rules with business process management (BPM) has been predicted for many years and is now a matter of fact. Every major BPM vendor has incorporated or acquired rules technology within their products and platforms. However, most rules offerings are only loosely integrated with processes at the task level. The use of business rules remains largely confined to managing isolated decisions services. Weak integration and isolation effectively relegates rules to an implementing role rather than a first class citizen in the capture and management of enterprise knowledge.

As the largest vendors bring their rules offerings to market and as standards from the W3C and OMG mature to adequacy, the opportunity for vendor-agnostic business rules management systems (BRMS) approaches. And continued improvement in end-user accessibility of BRMS promises ever less technical and ever more semantic expression and management of enterprise knowledge, including process and service models in addition to data models and business rules.

The ability to interchange more semantic models across major vendor offerings promises to dramatically increase the market demand for reusable, enterprise-relevant knowledge. But as knowledge becomes increasingly declarative and independent of implementations, it naturally becomes more ontological. Unfortunately, current ontological technologies are functionally inadequate from a business process or event processing perspective. These inadequacies include the lack of ontology for events, processes, states, actions, and other concepts that relate to change over time. Without such ontologies, rules or logic that govern processes or react to events must remain at the level of procedural implementation rather than declarative knowledge.

Until BRMS understand rules that refer to activities and events occurring within business processes, business rules applications will remain largely confined to discrete decisions, such as encapsulation within a decision service. By incorporating an adequate ontology of events and action, however, the knowledge management capabilities first developed in BRMS will broaden to encompass much of BPM and complex event processing (CEP). Given the fact that BRMS has been incorporated by the dominant platform vendors, modeling should move up from the relatively narrow perspective of a BRMS into the broader context of BPM and CEP.

The migration of business rules management into event and process contexts will emphasize the separation of business rules into ontology versus behavior. Modeling event-driven and business processes will correspond to defining ontology. Implementing event-driven and business processes will invoke behaviors at runtime. The ontology will be the same for BPM, CEP, or the BRMS, as will the behaviors. But the BRMS will know what is happening in terms of events and processes and it will know what processes it can invoke and what events it can signal.

As ontology management becomes increasingly relevant across application development, the limitations of related standards will also come into clearer focus. Commercial BRE are, for the most part, production rule systems that emphasize action over logic. This is best addressed by OMG's production rule representation (PRR) standard. But PRR is an isolated standard that

N. Bassiliades, G. Governatori, and A. Paschke (Eds.): RuleML 2008, LNCS 5321, pp. 4–5, 2008.
© Springer-Verlag Berlin Heidelberg 2008

includes no support for ontology or logic. W3C's web-ontology language (OWL) and rule interchange format (RIF) standards address ontology and logic, but not change or action. The same is true of OMG's SBVR, which emphasizes linguistics, in addition to ontology and logic, albeit in a way that remains disconnected with the OMG stack, including PRR.

Bridging events, processes and other aspects of reality that occur and change over time and incorporating action is a fundamental challenge for semantic technologies, especially formal logic, that must be addressed in a practical manner before rules standards and semantic technologies will bear substantial fruit in enterprise contexts.

Development and Verification of Rule Based Systems — A Survey of Developers

Valentin Zacharias

FZI Research Center for Information Technologies at the University of Karlsruhe
zach@fzi.de
http://www.fzi.de/ipe

Abstract. While there is great interest in rule based systems and their development, there is little data about the tools and methods used and the issues facing the development of these systems. To address this deficiency, this paper presents the results from a survey of developers of rule based systems.

The results from the survey give an overview of the methods and tools used for development and the major issues hindering the development of rule based systems. Recommendations for possible future research directions are presented.

The results point to verification and validation, debugging and overall tool support as the main issues negatively affecting the development of rule based systems. Further a lack of methodologies that appropriately support developers of these systems was found.

1 Introduction

With the application of rules on the Semantic Web, the continuing rise of business rule approaches and with initiatives to standardize and exchange rules, there is currently a renewed and increasing interest in rule based software systems and their development. At the same time, however, there is little data and overview about the way rule bases are used, developed and which challenges developers of these systems face; in sum there is little data that could be used to set priorities for research and development.

To address this shortcoming this paper presents a survey of the methods and tools used and the issues facing the development of rule based systems. Based on prior experience [1] and much older surveys (see related work) verification and particularly debugging issues were identified as particular important and the survey was designed to focus on these.

This paper starts with a short related work section introducing two much older studies that addressed similar questions, results from these survey are also cited throughout the text where similar questions were used. The next sections introduce the survey and presents data about the participants, their experience and the rule base systems they develop. The core results from the survey are then grouped into three sections, (1) Methods and Tools Used for Development, (2) Verification, Bugs and Debugging and (3) Issues and Comparison to Procedural

N. Bassiliades, G. Governatori, and A. Paschke (Eds.): RuleML 2008, LNCS 5321, pp. 6–16, 2008.

Programming. The conclusions highlight the most important findings and use these to derive possible directions for future research.

2 Related Work

In 1991 Hamilton et al. [2] questioned 70 people with respect to the state of the practice in knowledge based system verification and validation. The goal of the survey was to document the state of the practice. The insight gathered during the survey and follow up interviews was used to develop recommendations for further V&V research. The findings from this survey included that 62% of developers judged the developed expert system to be less accurate than the expert and 75% judged it to be less accurate than expected.

Also published in 1991 O'Leary [3] used a mailed survey to query 34 developers of knowledge based systems. This poll had the specific goal of testing a number of hypotheses with respect to prototyping of knowledge based systems. The core finding was that the prototyping development methodology is found to lead to more robust models and that it does not increase the validation difficulty.

In the broader world of *conventional* software engineering (i.e. using procedural or object oriented languages) Cusumano et al. [4] compared the practices used in software development across countries. Zhao and Elbaum [5,6] explored the use of quality assurance tools and techniques in open source projects. Finally Runeson et al. [7] summarize the (mostly experimental) empirical data with respect to quality assurance methods.

3 The Survey

The goal of the survey was to be an exploratory study of the methods and tools used, and the issues facing the developers of rule based systems. The survey focused on verification and in particular debugging as very important questions that in the authors experience are particularly problematic for the development of rule based systems. Some questions were also derived from specific hypotheses, described in detail below together with the results for the questions.

The survey is based on a very broad understanding of rule based systems, encompassing business rules, Prolog, Jess, SWRL and other languages. The author believes that in striving to be declarative program specifications, the use of an if-then structure and the use of an inference engine that automatically combines rules based on the task at hand; these systems share enough characteristics to be usefully grouped for the purpose of this survey.

The survey was designed to be answerable in less than 15 minutes; included 17 questions spread over three pages and was conducted using the SurveyMonkey [8] service. The survey with all questions in their original layout can be accessed at `http://vzach.de/papers/survey08.pdf`. Participants were asked to answer all questions with respect to the largest rule base in whose development they had been involved with in the past 5 years.

4 Participants

Participants were recruited though emails sent to public mailing lists concerned with rule based systems and mailing lists of academic institutes; invitations were also published on some blogs concerned with rule based systems. A chance to win a camera was given as additional incentive to motivate people to participate. 76 people opened the survey and 64 answered most questions; one reply was removed because it consisted of obviously nonsensical answers.

Because of the relatively modest number of participants this paper will constrain itself to reporting the results without attempting to find correlations between different questions of the survey; correlations which in any case would not attain statistical significance.

Table 1. Measures of the size of the rule base

	Mean	Median	Standard Deviation
Person Month Development			
PM for entire software	59	15	148
PM for rule base	9	5.5	15
Size of Rule Base			
Number of rules	1969	120	8693
Size of average rule	9.3	5	17
Size of largest rule	24	11	39
Developers involved			
Rule developers	3	2	4
Other software developers	3	1	8
Domain experts that created rules	1.5	1	2
Domain experts as consultants	1.9	1	2.5
Domain experts for V&V	1.7	1	2.4
Others	0.6	0	1.6

For the purpose of analysis the wide variety of systems used by the respondents was grouped into five groups:

- **Prolog:** 7 results; consisting of tuProlog(2), SWI Prolog (2), Visual Prolog (1), XSB (1) and Yap Prolog (1)
- **Declarative Rules:** 11 results; consisting of F-logic - ontoprise (3), SWRL - KAON2 (2), SWRL - Protege SWRL Tab (2), SWRL - PELLET (1), Jena Rules (1), WSML-Flight(1), Ontology Works Remote Knowledge Server (1) and Iris (1)
- **Business Rule Management Systems (BRMS):** 17 results; consisting of JBoss /Drools (8), Fair Isaac Blaze Advisor (3), Yasu Quick Rules (SAP) (2), BizTalk (1), NxBre (1), Acumen Business-Rule Manager(1), OpenRules (1) and Ilog JRules/.Net rules (1)

- **Shells:** 24 results, consisting of Jess (12), Clips (9), Mandarax (1), Jamocha (1) and KnowledgeWorks/LispWorks (1)
- **Other:** 1 results, using a 'Proprietary IT software engine'

The size of the reported systems varied widely (see Table 1); the average rule base consists of 2000 rules, has 9 conditions/body atoms per average rule and is developed in 9 person month. On average it is part of a much larger software system that takes almost 60 person months to develop. The largest system in the survey has 63000 (partly learned) rules is used for disease event analysis. The most time consuming took 100 person months to develop and is used to determine parameters to operate a medical imaging system. Slightly over 50% of the projects involve at least one domain expert that creates rules herself.

On average the people filling out the survey had 6.6 years of experience with rule based systems and 15 years experience with creating computer programs in general.

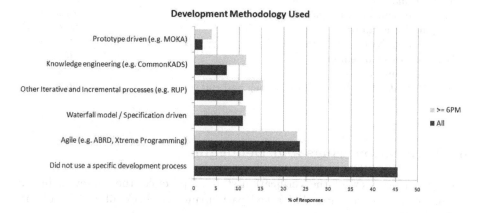

Fig. 1. The development methodology used, for all responses and for the 26 responses where the rule base development took at least 6 person months

The tasks of the rule bases (entered as free text) include workflow management, diagnosis, scoring, ontology reasoning, tutoring and planning. The rule bases are created for a multitude of different domains, including insurance, finance, health care, biology, computer games, travel and software engineering.

38% of the rule bases are commercially deployed, 26% are deployed in a research setting and 10% are in development with deployment planned. The remaining 26% are prototypes, 10% are prototypes that are also used by others than the developers.

5 Methods and Tools Used for Development

Participants of the survey were given a multiple choice question to describe the methods used for the development of the rule base. The answers (see figure 1)

showed that indeed a large part of rule based systems are created without any specific development process and that the rise of agile and iterative methods [9,10] is also visible for rule based systems. In 1991 Hamilton et al. [2] used a similar question and found that the most used model was the cyclic model (41%) and that 22% of the respondents followed no model[1].

The next questions asked participants for the tools used for the development. The results show that the most widely used tools for editing rule bases are still textual editors, with 33% and 28% of respondents stating that they use a simple text editor or a textual rule editor with syntax highlighting. With 26% of respondents using them, graphical rule editors are also widespread (see table 2).

Table 2. Use of tools for the development of rules

	Percent responses
Simple text editor	33%
Textual rule editor	28%
Constraint language, business language rule editor	10%
Graphical rule rditor	26%
Spreadsheets based rule editor	12%
Decision trees rule editor	9%
Rule Learning	5%
An IDE that allows to edit, load, debug and run rules	46%

6   Verification, Bugs and Debugging

To gain an overview of the verification state of practice, the survey included a multiple choice question that asked participants to check all V&V tools or methods that they use in their project.

The results show (see figure 2) that verification is dominated by testing (used by 90%) and code review (used by 78%). 74% of respondents do testing with actual data, 50% test with contrived data. Advanced methods of test organization are used by a minority, with only 31% doing regression testing and 19% doing structural testing with test coverage metrics. Code review is done equally by domain experts (53%) and developers (57%), most projects combine both (73% of projects doing code review do code review both by domain experts and developers). The system is used parallel to development in 17% of the projects; in 16% it is used by developers; in 14% by domain experts.

Leary [3] posed a similar question to the developers of expert systems in 1991, asking about the validation effort spend on particular methods. In the average over all responses he found that most effort is spend on testing with actual data (31% of validation effort), followed by testing with contrived data (17.9%), code

---

[1] However, care should be taken when comparing these numbers, since the sample of the surveys differs considerable.

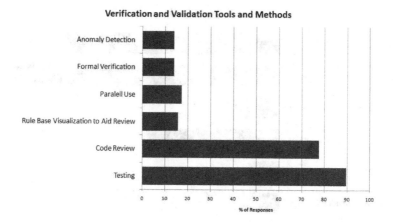

Fig. 2. The verification and validation methods and tools as percent of respondents that use a particular method

review by domain expert (17.6%), code review by developer (13%), parallel use by expert (12%) and parallel use of system by non-expert (7%).

6.1 Debugging Tools

Debugging is dominated by procedural debuggers, i.e. debuggers similar to the ones used in procedural programming; tools that allow to specify breakpoints, to suspend the inference engine and to explore the stepwise execution of the program[2]. 37% of the projects used a command line procedural debugger and 46% a graphical procedural debugger. Explanations are used by almost a quarter (23%) of the respondents for debugging.

Surprisingly widespread is the use of Algorithmic Debugging [12] and Why-Not Explanations (e.g. [13,14]), considering that to the best knowledge of the author there is no widely available and currently maintained implementation of either of these debugging paradigms. For the systems used by three of the five people that professed to use Algorithmic Debugging (JBoss rules/Drools and Fair Isaac Blaze Advisor) no mentioning of any such debugger could be found and it seems likely that the short explanation for this debugging paradigm given in the survey (*'system tries to identify error by asking user for results of subcomputations'*) was insufficient to convey the meaning of this concept. Similarly for Why-Not explanations three of the four respondents use systems for which such debuggers are not available. The remaining two responses for Algorithmic Debugging and the remaining one for Why-Not explanations use Prolog dialects, where such debuggers have existed/exist.

---

[2] An overview and description of the different debugging paradigms for rule based systems can be found in [11].

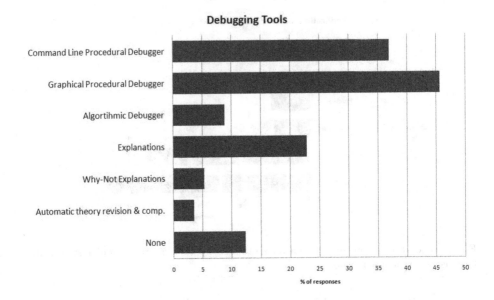

Fig. 3. Tools used for debugging

6.2 Bugs, Symptoms of Faults in the Rule Base

Debugging is the process of tracking down a fault based on error revealing values. The difficulty of this process and the kind of tools that can support it, depends on the error revealing values and how well these allow for the identification of the fault. In the authors experience based on F-logic [1] most faults in rule based systems cause a query to not return any result (the so called *no-result-case*) or to return less results than expected. This stands in contrast to the development with modern object oriented languages where in many cases at least a stack trace is available. This is problematic because a *no-result-case* gives only very little information for fault localization and means that most explanation approaches are not suitable for debugging (because these rely on the existence of a result). A question was included in the survey about the common symptoms of faults in the rule base to check whether this patterns of error revealing values holds for rule based systems in general.

The results (see table 3) show 'wrong results' as the most frequent bug, followed by 'no results' and 'partial results'. Most participants encounter not terminating tests and crashing rule engines only seldom. The results show also that 60% of participants frequently and 34% sometimes encounter a fault showing itself in the failure of the rule base to conclude an expected result/result part. For the developers of the system using *declarative rules* (see section 4), the no-result case is the most frequent bug. These results underline the need for debugging approaches to support users in diagnosing bugs based on missing conclusions.

Table 3. Bugs, symptoms of faults in the rule base in percent of responses. To 100% missing percent: respondents selected *not applicable.*

	frequent	seldom	never
A query/test would not terminate	7	57	30
A query/test did not return any result	38	47	11
A wrong result was returned	53	39	5
A part of the result missing	31	42	20
The rule engine crashed	9	47	38

7 Issues and Comparison to Procedural Programming

On the last page participants were asked to rank a number of possible issues as *Not an Issue, Annoyance* or *Hindered Development.* An average score was obtained for the issues by multiplying the *annoyance* answers with one, the *hinderance* answers with two and dividing by the number of all answers. The aggregated answers for this question are shown in the table below.

Table 4. Issues hindering the development of rule based systems. Numbers show the actual number of participants that selected an option. Please note that the 'Rule Expressivity' option was phrased in a way that asked also for things that could not easily be represented, not only things that could not be represented at all.

	Average	Not an Issue	Annoyance	Hindrance
Debugging	1	12	28	12
Determining completeness	0.76	18	27	6
Supporting tools missing/immature	0.67	26	17	9
Editing of rules	0.66	24	23	6
Determining test coverage	0.65	25	19	7
Inexperienced developers	0.58	31	13	9
Rule expressivity	0.5	33	12	7
Keeping rules base up to date	0.5	30	19	4
Understanding the rule base	0.47	31	19	3
Runtime performance	0.41	35	14	4
Organizing collaboration	0.41	35	14	4

The results show the issues of verification, validation and tool support as the most important ones. The issues of probably the largest academic interest - runtime performance and rule expressivity, are seen as lesser problem. This is particular interesting in the light of the fact that of the 7 survey participants that stated they were hindered by rule expressivity, none used a declarative rule system (for which these questions are debated the loudest).

These findings of verification and validation issues as the most important ones are similar to the finding of Leary [3]. He found that the *potentially biggest problems* were determining the completeness of the knowledge base and the difficulty to ensure that the knowledge in the system is correct.

In a final question participants were asked how a rule base development process compares to the development of a conventional program (created with procedural or object oriented languages) of similar size. A number of properties was given to be ranked with *Rule base superior, Comparable, Conventional program superior* and *Don't know*. The aggregated score for each property was determined by subtracting the number of *conventional program superior* answers from the *rule base superior* answers and dividing the result by the number of answers other than *don't know*.

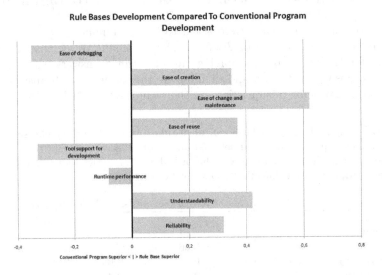

Fig. 4. Participants opinion about the strength and weaknesses of rule base development compared to that of 'conventional' programs. Positive numbers indicate that the majority thought rule bases to be superior, negative numbers that they thought conventional programs to be superior.

The participants of the survey judged rule bases to be superior in most respects. The largest consensus was that rule bases are indeed easier to change and to maintain. Ease of creation, ease of reuse, understandability and reliability are also seen as the strong points of rule based systems. A small majority saw conventional programs as superior in runtime performance; most saw rule bases as inferior in ease of debugging support and tool support for development.

8   Conclusion

The developers of rule based system are indeed seeing the advantages in ease of change, reuse and understandability that were expected from rule based (and declarative) knowledge and program specification. However, there are also issues hindering the development of these systems with verification and validation and the tool support for development topping the agenda.

Particularly debugging is seen most frequently as an issue negatively affecting the development of rule bases and as one area were rule base development is lacking behind procedural and object oriented program development. This relative difficulty of debugging may be caused either by a lack of refined debugging tools or by intrinsic properties of rule bases that make these hard to debug[3]; in either case it should be a priority topic for researchers and practitioners in the area of rule based systems. The results from the survey further show that many of the older innovative approaches for the debugging of rules are not used widely in practice; resuscitating these and refining them for practical use may be one way to tackle this debugging challenge of rule based systems[4]. That 60% of the projects frequently and another 34% sometimes encounter a fault showing itself in the failure of a rule base to conclude an expected result/result part show that the diagnosis of bugs based on missing conclusions must be a core feature of any new debugging tool.

Missing and immature tools support for the development of rule bases as the other big issue can be seen as another motivation for the already ongoing efforts to standardize rule languages. A tool supporting a standartized (and widely accepted) rule language could be developed with more resources, because its potential market would be larger than one for a tool supporting just one language in a fragmented rule landscape.

Finally this survey together with a literature research reveals a lack of methodological support for the development of rule bases: 38% of the larger projects in the survey use agile or iterative methods; at the same time there is no established agile or iterative methodology for rule based systems. Exploring, describing and supporting the transfer of agile and iterative methods to the development of rule based systems should be one major topic for the future development of rule based systems; notable first movements in this direction are the open sourcing of Ilogs Agile Business Rule Development methodology [15] and the authors own work on the adoption of eXtreme programming for rule development [16].

Acknowledgements

The author thanks Hans-Joerg Happel, Andreas Abecker, Michael Erdman, Katharina Siorpaes and Merce Mueller-Gorchs for their support in creating and realizing this survey.

References

1. Zacharias, V.: Rules as simple way to model knowledge: Closing the gap between promise and reality. In: Proceedings of the 10th International Conference on Enterprise Information Systems (to appear, 2008)

---

[3] [1] details some hypothesis on what those properties may be.

[4] See [11] for an overview including the older debugging approaches.

2. Hamilton, D., Kelley, K.: State-of-the-practice in knowledge-based system verification and validation. Expert Systems With Applications 3, 403–410 (1991)
3. O'Leary, D.: Design, Development and Validation of Expert Systems: A Survey of Developers, pp. 3–18. John Wiley & Sons Ltd., Chichester (1991)
4. Cusumano, M., MacCormack, A.C., Kemerer, F., Crandall, B.: Software development worldwide: The state of the practice. IEEE Software 20 (2003)
5. Zhao, L., Elbaum, S.: A survey on quality related activities in open source. SIGSOFT Software Engineering Notes 25(3), 54–57 (2000)
6. Zhao, L., Elbaum, S.: Quality assurance under the open source development model. Journal of Systems and Software 66, 65–75 (2003)
7. Runeson, P., Andersson, C., Thelin, T., Andrews, A., Berling, T.: What do we know about defect detection methods? IEEE Software 23, 82–90 (2006)
8. SurveyMonkey: Surveymonkey (2008) (accessed 2008-05-29),
   http://www.surveymonkey.com/
9. Larman, C., Basili, V.: Iterative and incremental development: A brief history. IEEE Computer, 47–56 (June 2003)
10. MacCormack, A.: Product-development practices that work. MIT Sloan Management Review, 75–84 (2001)
11. Zacharias, V.: The debugging of rule bases. In: Handbook of Research on Emerging Rule-Based Languages and Technologies: Open Solutions and Approaches. IGI Global (to appear, 2009)
12. Shapiro, E.Y.: Algorithmic program debugging. PhD thesis, Yale University (1982)
13. Chalupsky, H., Russ, T.: Whynot: Debugging failed queries in large knowledge bases. In: Proceedings of the Fourteenth Innovative Applications of Artificial Intelligence Conference (IAAI 2002), pp. 870–877 (2002)
14. Becker, M., Nanz, S.: The role of abduction in declarative authorization policies. In: Proceedings of the 10th International Symposium on Practical Aspects of Declarative Languages (PADL) (2008)
15. Ilog: Agile business rule development (2008) (accessed 2008-05-31),
   http://www.ilog.com/brms/media/ABRD/
16. Zacharias, V.: The agile development of rule bases. In: Proceeedings of the 16th International Conference on Information Systems Development (2007)

Connecting Legacy Code, Business Rules and Documentation

Erik Putrycz and Anatol W. Kark

National Research Council Canada
{erik.putrycz,anatol.kark}@nrc-cnrc.gc.ca

Abstract. By using several reverse engineering tools and techniques, it is possible to extract business rules from legacy source code that are easy to understand by the non-IT experts. To make this information usable to business analysts, it is necessary to connect the artifacts extracted to existing documents. In this paper, we present how we use source code analysis and keyphrase extraction techniques to connect legacy code, business rules and documentation.

Keywords: Reverse engineering, Business Rules, Information Retrieval, System modernization, Keyword Extraction.

1   Introduction

Governments and large corporations maintain huge amount of the legacy software as part of their IT infrastructure. In 2008, 490 companies of the Fortune 500 are still using legacy systems to process more than 30 billion transactions or \$1 trillion worth of business each and every day. In Canada, a recent report [1] estimates that 60,000 employees - which is 10% of the 600,000 total ICT employment - are working with legacy systems.

Understanding and discovery of business rules play a major role in the maintenance and modernization of legacy software systems. The recovery of business rules supports legacy asset preservation, business model optimization and forward engineering [1]. According to a recent survey [2], about half of the companies who reported difficulties in modernizing their legacy systems, said that a major issue was the fact that "hard-coded and closed business rules" make it difficult to adapt their systems to new requirements and migrate to more modern environments. In our previous work [3], we described how we extract business rules from legacy code. However, business analysts need more than just the artifacts from code. In this paper, we present how we use HTML extraction techniques and keyphrase analysis to link the source code artifacts to their technical and other related documents. We also describe an intuitive navigation tool designed for the use by the business analysts. This paper presents the results of analyzing approximately 700,000 lines of the COBOL source code and 4000 HTML and Microsoft Word documents.

N. Bassiliades, G. Governatori, and A. Paschke (Eds.): RuleML 2008, LNCS 5321, pp. 17–30, 2008.
© Springer-Verlag Berlin Heidelberg 2008

2   Background

In this paper, we focus on large legacy systems. These systems share the following characteristics [4]:

- Often run on obsolete hardware that is slow and expensive to maintain.
- Are expensive to maintain, because documentation and understanding of system details is often lacking and tracing faults is costly and time consuming.
- Lack of clean interfaces make the integration with other systems difficult.
- Are difficult, if not impossible, to extend.

These systems are found in most sectors such as pay, insurance, warehouse logistics and many other sectors.

Several factors motivate industry and governments to migrate from their legacy systems [1]:

- The hardware is no longer supported (e.g. chip sets become obsolete);
- The system becomes error prone;
- The system or applications no longer fit the business;
- Key in-house people retire or leave and replacements are difficult to recruit and/or train (including contractors);
- There is a herd mentality in some industries toward a new technology solution;
- Senior management becomes concerned about the risks involved.

These legacy systems often date from 1970s when the concepts of proper software engineering were relatively new and proper documentation techniques were not a concern. Old COBOL applications are typically monolithic, with hundreds of thousands of lines of code mixing the business logic, user interface, and transaction data in no particular sequence [5]. As a consequence, migrating from these systems is a complex and expensive process. When possible, these legacy system are kept intact, wrapped and integrated in a more modern system. In other cases, it is necessary to re-engineer the whole system.

Without the knowledge from the legacy code, analysts would need to rewrite all the business rules using all the legislations, policies and other agreements. These rules usually consists of calculations, exceptions and other elements that can be recovered from the code.

In our previous publication [3], we detailed how we extract business rules from legacy COBOL code and provide a basic linear navigation. The feedback from analysts showed us that this linear navigation is not sufficient and they need to locate business rules for a specific topic or an existing document.

In the rest of the paper, we detail how we achieved this objective. First, we present two existing solutions and detail why they are not appropriate for our case (Section 3). Then, we present our approach, which consists of first extracting business rules and other artifacts from source code and building a knowledge database (Section 4). Once that is built, we link the identifiers used in business rules to existing technical documents (Section 5). Using these links and keyphrase extraction techniques, we are able to connect the business rules extracted from the source code to many other documents (Section 6). In Section 7 we present short conclusions and outline further work.

3 Related Work

Several approaches exist to connect documents to source code. Witte et al. [6,7] propose an approach to extract semantic information from source code and documents and populate an ontology using the results. They use native language processing to analyze technical documentation and identify code elements in the documentation. To connect the documents together, they use a fuzzy set theory-based coreference resolution system for grouping entities. Rule-based relation detection is used to relate the entities detected. The text mining approach achieves 90% precision in the named entities detection. However after adding the source code analysis, the precision drops to 67%.

In our work, we use only data description documents. Since the those documents were not by and large "written in prose" the information contained these documents had to be extracted using different techniques from natural language processing. Additionally, our approach is centered on the code and documents used to support artifacts extracted from the code.

In [8], Antoniol et al. look at the problem of recovering traceability links between the source code of a system and its free text documentation. They use information retrieval techniques to trace C++ code to free text documents. The process normalizes identifiers from the source code and does a morphological analysis of the software documents. Both are indexed and a classifier connects them. Unfortunately the precision achieved isn't very high.

In our case, the documentation considered is quite different. We focus our work on the application specific data instead of the programming interface. The data considered do have complex names that can be found accurately with full text search. Thus, natural language analysis is not necessary to locate identifiers. In addition, since the documents considered share a similar structure, it is possible to locate specific parts of the documents their formatting.

4 Extracting Knowledge from Source Code

In the context of a system modernization, our objective is to provide tools that extract business rules from the source code, to support both a legacy system maintenance and new system construction. Consequently, the requirements for extraction tools are:

- All extracted artifacts have to be traceable to the source code;
- Business rules must be at high level and understandable by all stakeholders involved, including business analysts.

In this section, we focus on analyzing and extracting business rules from the COBOL programming language in a way that satisfies our stated objectives. The process and tools presented in this paper have been implemented for COBOL and were developed as part of two large system modernization projects in which we are currently involved. While COBOL is our current source language, we believe that the same method can be applied to other programming languages.

4.1 Business Rules Definition

We use a small subset of SBVR [9] called "production business rules". They have the following syntax:

< conditions >< actions >

where < conditions >

1. consists of one or more Boolean expressions joined by logical operators ("and", "or", "not")
2. must evaluate to a "true" state for the rules actions to be considered for execution;

   and < actions >

1. consists of one or more Action expressions
2. requires the rule conditions to be satisfied (evaluate to true) before executing.

4.2 From COBOL to Business Rules

The structures of legacy COBOL and todays object oriented programs are radically different. COBOL has a limited and simple structure - each program contains a sequence of statements grouped into paragraphs which are executed sequentially. COBOL has only two forms of branching; one to execute external programs (CALL) and another to transfer the control flow to a paragraph (PERFORM). PERFORM is also used for iterations. Each program is associated with a single source file.

To extract useful information from the source code, it is necessary to find heuristics that separate setups and data transfers (usually from flat files) from business processing. To separate those two aspects, we focus on single statements that carry a business meaning such as calculations or branching since they most often represent high level processing.

In the case of branching (as defined above), Paragraph names and external programs can be either traced to documentation or the names themselves can possibly be self-explanatory. In code we have investigated, we found large number of paragraph names such as "CALCULATE-BASIC-PAY" which can be understood by a business analyst.

Once calculations and branching are located, it is necessary to construct their context. By context we mean all conditions in which a calculation or branching operation happens. This includes all IF statements under which an operation might be executed plus iteration information. We consider two contexts for one operation.

- local context: the IF statement surrounding the operation;
- global context: all the possible IF statements - starting from the beginning of execution - that can lead to the operation.

Depending on the operation and on the information required, either the local or the global context might be the most appropriate.

4.3 Knowledge Extraction Process

This business rules construction process is based on abstract syntax tree (AST) analysis. An abstract syntax tree is a tree, where the internal nodes are labeled by operators, and the leaf nodes represent the operands of the operators. This tree is obtained by parsing the source code. The parser required to construct the AST for this task had to be specifically designed to analyze and include all elements of the source code including comments. Many existing parsers are designed for other purposes such as syntax verification or transformation and they did not suit our objectives.

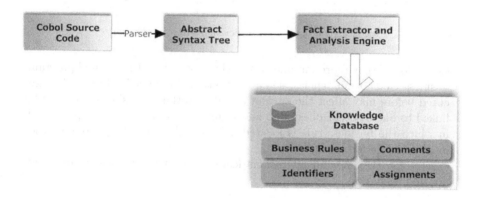

Fig. 1. Extraction step 1: Source code analysis

The business rules extraction is divided into two steps. (Figure 1). First, we analyze the AST and construct a knowledge database. Second, once that database is constructed, we simplify the data collected and link the artifacts, wherever possible, to the existing documentation (Section 6).

The extracted elements of the knowledge database are:

- production business rules: if condition, do action;
- conditions: one or more Boolean expressions with variables and constants joined by logical operators;
- business rules: an action, possibly a condition and a comment from the source code; where action can either be branching statement or calculations with identifiers and constants;
- identifiers: variables used in conditions and calculations;

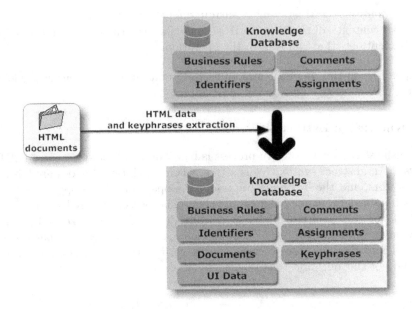

Fig. 2. Extraction step 2: Documentation integration

- Code blocks: they represent one or more lines of code in the original program;
- business rules dependencies: some calculations in other business rules executed before may affect the current calculation thus a business rules can be linked to one or other ones; loop and branching information: the paragraph in which the business rule is located may be called in a loop from another location in the program;
- exceptions: they are all operations leading to an error message or to an aborted execution.

The second step (Figure 2), connects the documentation to the artifacts previously extracted (detailed in Section 6). All documents are first converted to HTML format and then keyphrases are extracted. All this information is added to the database. In addition, all the navigation indexes used on the end-user graphical interface are calculated and added to the database.

4.4   Identifying Temporary Identifiers

A common pattern found in legacy programs is the usage of temporary identifiers for performing a set of operations, instead of the identifiers attached to the data fields. An example is presented in Figure 3. To link the identifiers to a data field in many calculations, it is necessary to track assignments and other operations that copy the value of one identifier into another. Tracing these temporary identifiers can be very valuable for documenting the operations being performed. When the identifers of operation are documented through temporary identifers, we call a *transitive connections* the connection between the document and the identifiers.

```
Load identifier A from database
Temporary identifier Ta = A
. . .
Calculate Ta = . . .
. . .
Set A = Ta
Save A in database
```

Fig. 3. Temporary identifier usage example

This connection is not totally accurate because we cannot verify the order of the assignations through the execution. The accuracy of the translations is discussed in Section 5.2.

5 Linking Identifiers to Technical Documents

In our previous work [3], we used a translation table and a list of abbreviations to convert an identifier to a meaningful name. However, that approach did not provide satisfactory results with a different set of COBOL programs. In addition to this issue, the accuracy of the results obtained with the translation table and abbreviations was dependent on the accuracy of the translation tables themselves. Those tables were also "legacy" and were proven not well maintained.

5.1 Translating Identifiers

To translate identifiers to business names, we decided to rely on an existing documentation of the data. The documentation describes all data fields used in programs and in databases. An example of a document is showed in Figure 4. (The actual data was obfuscated.)

The document details a data element and its use across all the systems. All these documents have been generated by technical writers and they all share a similar structure with sections and fields.

We make use of this similarity to extract information. As first step, all documents are normalized to a hierarchical structure. In our knowledge database, documents have the following structure:

- Document group: contains a set of documents;
- Document: contains a sequence of sections;
- Section: contains a sequence of paragraphs.

In order to use the documentation to translate a COBOL identifier into a term understood by business analysts , it is necessary to analyze the document, locate the identifier and the context in which it is found. Locating the identifier in the documents has been achieved using a customized full text search engine. Full text search engines usually tokenize all text using a pre-defined set of separators. We

```
Axx Indicator

Technical Name:AXX_YYY_IND
Definition: AXX YYZZs Indicator within YYZZs Codes Control File
Model Status:System Information/Skip: Reviewed during conceptual data modeling process
- bypassed because this is a current computer system/application specific term
and may not be required
for a system rewrite/business re-engineering.
Indicates whether a particular YYZZs can appear in an AXX transaction (deduction and YYZZs form).
System(s):System1
System2
System3
Element Type:Business
Data Class:Indicator
Data Type:Base
Data Structure:1 character, alphanumeric


System1
Notes:Synonym is: OL-YYZZ-AXX
V1-YYZZ-AXX
V2-YYZZ-AXX
Valid Values:Y, N
Input Forms:N/A
Element Name:YYZZ-AXX
- subordinate to:GO-YYZZ
GO-YYZZSES
GO-DATA
GOSS
Picture:PIC X(01)
Subordinate
Elements:N/A
File ID/Records
Description
MM200-XXXX-YYYY-LR logical record used by input/output module
MM401-SB-XXXX-YYYY-MMMMMM logical record used to build online screen


System2
Notes:Synonym is: OL-YYZZ-AXX
V1-YYZZ-AXX
V2-YYZZ-AXX
Valid Values:Y, N
Input Forms:N/A
Element Name:YYZZ-AXX
- subordinate to:GO-YYZZ
GO-YYZZSES
GO-DATA
GOSS
Picture:PIC X(01)
Subordinate
Elements:N/A
File ID/RecordsDescription
MM200-AAAA-BBBB-LR logical record used by input/output module
FF401-XX-YYYY-ZZZZ-UUUUUU logical record used to build online screen

System3
Notes:Synonym is: OL-YYZZ-AXX
V1-YYZZ-AXX
V2-YYZZ-AXX
Valid Values:Y, N
Input Forms:N/A
Element Name:YYZZ-AXX
- subordinate to:GO-YYZZ
GO-YYZZSES
GO-DATA
GOSS
Picture:PIC X(01)
Subordinate
Elements:N/A
File ID/Records Description
MM200-ZZXX-MMNN-LR logical record used by input/output module
MM401-ASDFG database record
MM401-AA-BBBB-CCC logical record used to build online screen
```

Fig. 4. Sample technical document

had to implement a specific tokenizer that filters identifiers in the documents and leaves them intact in the index. In the documentation, inside each section, are a list of field definitions in the form *"field: value"*. Once an identifier is located in a section, we verify that it is located in a field value and that the section name and the field name are valid . Locating valid positions for identifiers is done in three steps

1. Extract all field names using the HTML markup and position in HTML tree;
2. Consolidate and normalize the extracted field names and remove false positives (because the documents are hand written, several variants exist for each field name and the HTML markup is not used consistently);
3. Re-parse all the HTML documents with the normalized field names and extract correct sections and fields.

If the location is valid, the title of document is used as translation of the technical name. For instance, when looking for a translation of *MM200-ZZXX-MMNN-LR* in Figure 4, we locate it in section "System 3" and in the field "File ID/Records Description".

Despite the fact that legacy system are usually poorly documented, we expect that similar documents such as the ones used in this case and presented in Figure 4 can be found. A technical documentation on a system is not vital for running an IT system, however understanding the data processed is necessary.

5.2   Accuracy of Translations

With the process described above, when translating an identifier found only once and in a valid location, we can assume the translation is accurate. When an identifier is found in multiple documents, we introduce for each identifier translation an accuracy factor called tr_a. tr_a is calculated as: $tr_a = 1/n_{documents}$. In addition, when a transitive connection is involved, because different execution paths can lead to different results, we decided to decrease the original accuracy. Currently we use 70% of the original accuracy. It is an arbitrary number used to decrease the accuracy when transitive connections are involved, this number has to be tuned and balanced with the original calculation of tr_a.

To achieve the best accuracy, our implementation calculates all possible translations and uses the one with the highest accuracy. In addition, on the user interface, we offer a configuration setting to specify the accuracy level in order to eliminate long transitive connection paths.

5.3   Extracting Connections between Documents

Once the identifiers have been linked to data documentation, it is possible to link the documents to business rules. To link a document to a business rule, we locate all identifiers that are referenced in a document. Then, we search for all business rules that use one of the identifiers either as condition or as calculation and we link them to the document.

This offers a first level of navigation for locating any operation in a program connected to a data field.

To help analysts build a data flow in a new system and verify their results, we can extract dependencies between documents based on the business rules through a whole system. We consider two documents connected when at least one identifier found in each document appears in a business rule.

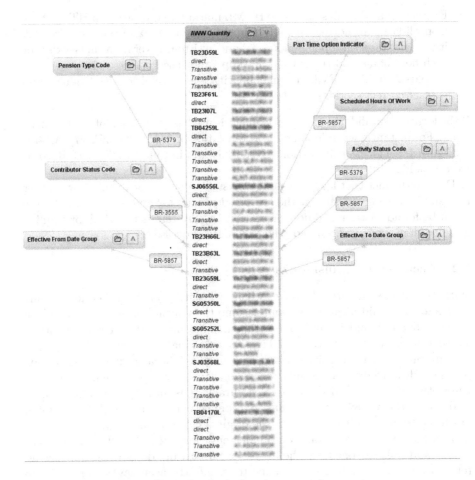

Fig. 5. Example of relations betweens documents

Figure 5 shows relations extracted from one document (called "AWW Quantity") and shows all other data sources involving *AWW Quantity* in an operation. A connector between two documents means that they are connected through identifiers in the business rule. The label on each connector is a business rule. The details on *"AWW Quantity"* are all programs and all identifiers connected to this documents and used in business rules.

6   Connecting External Documents to Code with Keyphrases

In the previous section, we built a navigation with identifiers and documents. However, this navigation is still at a low level, and a higher level navigation with business rules and additional text documents is necessary. A typical scenario is to search for all business rules related to an external text document.

Using the identifier translations established previously, we are able to show all business rules related to a data field and its documentation. To improve the navigation, we decided to use keyphrase extraction to index all documents including the data documentation (that is being used for translating identifiers). This provides another benefit, an additional navigation with keyphrases and topics.

Indexing documents with keyphrases is done using a technique called *keyphrase* extraction detailed below.

6.1   What Is Keyphrase Extraction

According to [10,11], a keyphrase list is a short list of phrases (typically five to fifteen noun phrases) that capture the main topics discussed in a given document. They are often referred as keywords.

There are two main algorithms for keyword extraction, Kea [12,13] and *extractor* [10,11]. Both algorithms share the same basic techniques and measures, but Kea is based on naive Bayes whereas Extractor relies on a genetic algorithm - GenEx, adapted for its purposes. For our work we used Kea to validate the usage of keyphrase extraction because of its easy to use interface.

6.2   Kea

For each candidate phrase Kea computes four feature values:

- **TFxIDF** is a measure describing the specificity of a term for this document under consideration, compared to all other documents in the corpus. Candidate phrases that have high TFxIDF value are more likely to be keyphrases.
- **First occurrence** is computed as the percentage of the document preceeding the first occurrence of the term in the document. Terms that tend to appear at the start or at the end of a document are more likely to be keyphrases.
- **Length** of a phrase is the number of its component words. Two-word phrases are usually preferred by human indexers.
- **Node degree** of a candidate phrase is the number of phrases in the candidate set that are semantically related to this phrase. This is computed with the help of the thesaurus. Phrases with high degree are more likely to be keyphrases.

Kea comes out of the box without any training data, so it is necessary to provide a training data containing documents and manually extracted keyword so that it can build a model. The model contains the four feature values described in the previous section.

We used for training a set of 341 various types documents with manually specified keyphrases. A similar set has been used for *extractor* and has been proven sufficient to train the model for any kind of document. Given the feature values used, the model generated is then generic enough for any domain.

6.3   Results

To evaluate the grouping of document with keyphrase extraction, we used two set of documents. The first set of documents are the data field documents used previously to translate identifiers (detailed in Section 5). This set consists of 3603 files. The second set consists of 352 files which are documents frequently used by business analysts to lookup information on business rules.

Results of keyphrase extraction are presented in table 1. The keyphrase extraction tools work on a per document basis, so the keywords extracted are not consistent through the whole set. A total of 5608 unique keywords has been found. 4106 keywords (73%) have only one document matched in each set and thus are not useful for grouping documents.

To connect other documents with source code, we evaluated how many documents from set 1 are connected with the set 2. We found out that 329 documents in the set 1 (93%) are connected with 1941 in set 2 (53%) through 156 keyphrases. These connections may be sufficient and may require manual work to improve the accuracy but without any additional semantic information on the documents, the possibilities are limited.

The keyphrases found have not yet been validated by business analysts, however since most of them appear in their glossary, we consider them relevant.

Table 1. Keyphrase extraction statistics

Total number of documents in set 1	352
Total number of documents in set 2	3603
Total number of keyphrases	5608
Number of keyphrases with at least one document matched in each set	156
Number of keyphrases with at least two documents matched in set 1	207
Number of keyphrases with at least two documents matched in set 2	1427
Number of keyphrases with only one document matched in set 1	359
Number of keyphrases with only one document matched in set 1	3771

7   Conclusions

This paper presents means of helping a modernization process by extracting business rules from legacy source code and connecting the business rules to existing documents. The novelty in this research is that reverse engineering, information extraction and natural text processing are used to connect the code to documents. This enables to translate business rules into non-technical terms and helps business analysts to locate business rules using existing documents. The extraction of rules from source code is divided into two main steps. First, the source code is parsed into an abstract syntax tree to locate the calculations and other elements of business rules. Second, a context is constructed for this information to transform it into a form which is easier to understand. We translate identifiers using data documentation and by analyzing the usage of temporary identifiers. Once documentation is linked to identifiers, keyphrase extraction techniques enables to connect external documents.

This process and tools are currently being used in one large modernization project where business rules play a critical role. Feedback from business analysts has been extremely positive about the value and understandability of the information extracted. We are currently planning to improve document analysis to add more sources of documentation for identifiers. Also, we are looking at improving the context of the business rules and simplifying the results found.

Acknowledgements

We would like to thank to Peter Turney for all his help on keyphrase extraction.

References

1. Senik, D.: Doyletech Corporation, Associates Inc.: Legacy applications trend report. Technical report, Information and Communications Technology Council (May 2008)
2. Software AG: Customer survey report: Legacy modernization. Technical report (2007)
3. Putrycz, E., Kark, A.: Recovering business rules from legacy source code for system modernization. In: Paschke, A., Biletskiy, Y. (eds.) RuleML 2007. LNCS, vol. 4824, pp. 107–118. Springer, Heidelberg (2007)
4. Bisbal, J., Lawless, D., Wu, B., Grimson, J.: Legacy information systems: issues and directions. Software, IEEE 16(5), 103–111 (1999)
5. Ricadela, A., Babcock, C.: Taming the beast. InformationWeek (2003)
6. Witte, R., Li, Q., Zhang, Y., Rilling, J.: Ontological text mining of software documents. In: Kedad, Z., Lammari, N., Métais, E., Meziane, F., Rezgui, Y. (eds.) NLDB 2007. LNCS, vol. 4592, pp. 168–180. Springer, Heidelberg (2007)
7. Witte, R., Zhang, Y., Rilling, J.: Empowering software maintainers with semantic web technologies. In: Proceedings of the 4th European Semantic Web Conference (2007)
8. Antoniol, G., Antoniol, G., Canfora, G., Casazza, G., De Lucia, A.: Information retrieval models for recovering traceability links between code and documentation. In: Canfora, G. (ed.) Proc. International Conference on Software Maintenance, pp. 40–49 (2000)
9. OMG: Semantics of Business Vocabulary and Business Rules (SBVR). dtc/06-08-05

10. Turney, P.D.: Learning algorithms for keyphrase extraction. Information Retrieval (2), 303–336 (2000)
11. Turney, P.: Learning to extract keyphrases from text (1999)
12. Frank, E., Paynter, G.W., Witten, I.H., Gutwin, C., Nevill-manning, C.G.: Domain-specific keyphrase extraction (1999)
13. Medelyan, O., Medelyan, O., Witten, I.: Thesaurus based automatic keyphrase indexing. In: Witten, I. (ed.) Proc. 6th ACM/IEEE-CS Joint Conference on Digital Libraries JCDL 2006, pp. 296–297 (2006)

Verifying Resource Requirements for Distributed Rule-Based Systems

Natasha Alechina, Brian Logan, Nguyen Hoang Nga, and Abdur Rakib*

University of Nottingham, Nottingham, UK
{nza,bsl,hnn,rza}@cs.nott.ac.uk

Abstract. Rule-based systems are rapidly becoming an important component of 'mainstream' computing technologies, for example in business process modelling, the semantic web, sensor networks etc. However, while rules provide a flexible way of implementing such systems, the resulting system behaviour and the resources required to realise it can be difficult to predict. In this paper we consider the verification of system behaviour and *resource requirements* for distributed rule-based systems. More specifically, we consider distributed problem-solving in systems of communicating rule-based systems, and ask how much time (measured as the number of rule firings) and message exchanges does it take the system to find a solution. We show how standard model-checking technology can be used to verify resource requirements for such systems, and present preliminary results which highlight complex tradeoffs between time and communication bounds.

1 Introduction

Rule-based approaches offer significant advantages to the application developer: their focus on the declarative representation of small, relatively independent, knowledge units makes it easier for developers and even end users to rapidly develop and maintain applications — in many cases the information required to develop the application is already codified in terms of rules expressed in natural language, e.g., describing a business process.

However, while the adoption of rule-based approaches brings great benefits in terms of rapid development and ease of maintenance, they also present new challenges to application developers, namely how to ensure the *correctness* of rule-based designs (will a rule-based system produce the correct output for all legal inputs), *termination* (will a rule-based system produce an output at all) and *response time* (how much computation will a rule-based system have to do before it generates an output).

These problems become even more challenging in the case of *distributed rule-based systems*, where the system being designed or analysed consists of several communicating rule-based programs which exchange information via messages, e.g., a semantic web application or a sensor network. A communicated fact (or sensor reading) may be added asynchronously to the state of a RBS while the system is running, potentially

---

* This work was supported by the Engineering and Physical Sciences Research Council [grant number EP/E031226].

N. Bassiliades, G. Governatori, and A. Paschke (Eds.): RuleML 2008, LNCS 5321, pp. 31–38, 2008.

triggering a new strand of computation which executes in parallel with current processing. To be able to provide response time guarantees for such systems, it is important to know how long each rule-based system's reasoning is going to take. In other situations, for example a rule-based system running on a PDA or other mobile device, the number of messages exchanged may be a critical factor.

Verifying properties such as correctness, termination and resource requirements of rule-based systems is extremely challenging. Ironically, the very features which make rule-based systems attractive from a development point of view—the separation of the application logic and execution engine and the ease with which rules can be added or modified (often by end users)—make it hard to predict the overall behaviour of the system or the implications of changing a particular rule. In this paper, we present a framework for the automated verification of time and communication requirements in distributed rule-based systems. We consider distributed problem-solving in systems of communicating rule-based systems, and ask how much time (measured as the number of rule firings) and message exchanges does it take the system to find a solution. We show how standard model-checking technology can be used to solve such problems. Using simple examples, we show how the Mocha model checker [1] can be used to analyse trade-offs between time and communication bounds in a distributed rule-based system.

The structure of the paper is as follows. In section 2 we introduce a simple model of the kinds of distributed rule-based system we want to verify. We describe the encoding of such systems in the input language of Mocha model-checker in section 3. Model-checking experiments are described in section 4. We discuss related work in 5 and conclude in section 6.

2 Distributed Rule-Based Systems

In this section, we introduce a model of a distributed rule-based system and the measures of time and communication resources required to solve a distributed reasoning problem.

We assume that the system consists of n individual rule-based systems or nodes, where $n \geq 1$. Each node is identified by a value in $\{1, \ldots, n\}$, and we use variables i and j over $\{1, \ldots, n\}$ to refer to nodes. Each node i has a *program*, consisting of propositional Horn clause rules, and a working memory, which contains facts (propositions). The restriction to propositional rules is not critical: if the rules do not contain functional symbols and we can assume a fixed finite set of constant symbols, then any set of first-order Horn clauses and facts can be encoded as propositional formulas. If a node i has a rule $A_1, \ldots, A_n \rightarrow B$, the facts $A_1, \ldots, A_n$ are in i's working memory and B is not in i's working memory in state s, then i can fire the rule, adding B to i's working memory in the successor state s'.

In addition to firing rules, nodes can exchange messages regarding facts currently in their working memory. The exchange of information between nodes is modelled as an abstract *Copy* operation: if a fact A is in node i's working memory in state s and A is not in the working memory of node j, then in the successor state s', A can be added to node j's working memory. Intuitively, this corresponds to the following operations

Time	Node 1	Node 2
t_0	$\{A_1, A_2, A_3, A_4\}$	$\{A_5, A_6, A_7, A_8\}$
operation:	RuleB2	RuleB4
t_1	$\{A_1, A_2, A_3, A_4, B_2\}$	$\{A_5, A_6, A_7, A_8, B_4\}$
operation:	RuleB1	RuleB3
t_2	$\{A_1, A_2, A_3, A_4, B_1, B_2\}$	$\{A_5, A_6, A_7, A_8, B_3, B_4\}$
operation:	RuleC1	RuleC2
t_3	$\{A_1, A_2, A_3, A_4, B_1, B_2, C_1\}$	$\{A_5, A_6, A_7, A_8, B_3, B_4, C_2\}$
operation:	Idle	Copy (C_1 from node 1)
t_4	$\{A_1, A_2, A_3, A_4, B_1, B_2, C_1\}$	$\{A_5, A_6, A_7, A_8, B_3, B_4, C_1, C_2\}$
operation:	Idle	RuleD1
t_5	$\{A_1, A_2, A_3, A_4, B_1, B_2, C_1\}$	$\{A_5, A_6, A_7, A_8, B_3, B_4, C_1, C_2, D_1\}$

Fig. 1. Example 1

rolled into one: j asking i for A, and i sending A to j. We assume copy operations are guaranteed to succeed and take one tick of system time. A node can also perform an Idle operation (do nothing).

A problem is considered to be solved if one of the nodes has derived the goal. The time taken to solve the problem is taken to be the total number of steps by the whole system (nodes firing their rules or copying facts in parallel, at most one operation executed by each node at every step). This abstracts away from the cost of rule matching etc. This assumption is made for simplicity and a single 'tick' can be replaced with a numerical value reflecting real time taken by the system to fire a rule (worst case or average). The amount of communication required to solve the problem is taken to be the total number of copy operations performed by all nodes. Note that the only node which incurs the communication cost is the node which performs the copy. As with our model of time, the assumptions regarding communication are made for simplicity; it is straightforward to modify the definition of communication so that, e.g., the 'cost' of communication is paid by both nodes, communication takes more than one tick of time, and communication is non-deterministic.

The execution of a distributed rule-based system can be modelled as a state transition system where states correspond to combined states of nodes (set of facts in each node's working memory) and transitions correspond to nodes performing actions in parallel, where each node's action is either a single rule firing, a copy action, or an idle action.

As an example, consider a system of two nodes, 1 and 2. The nodes share the same set of rules:

$$\textbf{RuleB1 } A_1, A_2 \rightarrow B_1 \quad \textbf{RuleB2 } A_3, A_4 \rightarrow B_2$$

$$\textbf{RuleB3 } A_5, A_6 \rightarrow B_3 \quad \textbf{RuleB4 } A_7, A_8 \rightarrow B_4$$

$$\textbf{RuleC1 } B_1, B_2 \rightarrow C_1 \quad \textbf{RuleC2 } B_3, B_4 \rightarrow C_2$$

$$\textbf{RuleD1 } C_1, C_2 \rightarrow D_1$$

The goal is to derive D_1. Figure 1 gives a simple example of a run of the system starting from a state where node 1 has A_1, A_2, A_3 and A_4 in its working memory, and node 2 has A_5, A_6, A_7, A_8. In this example, the nodes require one copy operation and five time steps to derive the goal. (In fact, this is an optimal use of resources for this problem, as verified using model-checking, see section 4).

Throughout the paper, we will use variations on this synthetic 'binary tree' problem, in which the A_is are the leaves and the goal is the root of the tree, as examples. We vary the number of rules and the distribution of 'leaf' facts between the nodes. For example, a larger system can be generated using 16 'leaf' facts $A_1, \ldots, A_{16}$, adding extra rules to derive B_5 from A_9 and A_{10}, etc., and a new goal E_1 derivable from D_1 and D_2. We will refer to this as a '16 leaf example'. We have chosen this sample problem because it is typical of a class of distributed reasoning problems and can be easily parameterised by the number of leaf facts and the distribution of facts and rules among the nodes.

3   Model-Checking Resource Requirements

We are interested in verifying properties of the form 'if the facts $A_1, \ldots, A_n$ are assigned to the nodes of a distributed rule-based system in a particular way, the system will (or will not) conclude Q in less than t timesteps and fewer than m messages'. In general it is impractical to run the system and count steps and messages for all possible interactions between the nodes to establish such properties. What is required is some automated method of verifying such properties which considers all possible system traces.

In this section, we show how the transition system representing a distributed rule-based system can be encoded as an input to a model-checker to allow the automatic verification of the properties expressing resource bounds. Model checking is an automated verification procedure in which the system to be verified is represented by a (finite) model M for an appropriate logic, the property to be verified is represented by a formula ϕ in the same logic, and the verification consists in computing whether M satisfies ϕ [2]. Originally developed for hardware verification, it is increasingly being applied to the verification of complex software systems. For the experiments reported here, we have used the Mocha model checker [1], due to the ease with which we can specify a system of concurrently executing communicating rule-based systems in *reactive modules*, the description language used by Mocha.

In Mocha, the state of the system is described by a set of *state variables* and each system state corresponds to an assignment of values to the variables. The presence or absence of each fact in the working memory of a node is represented by a boolean state variable $n_i A_j$ which encodes node i's belief in fact A_j. The initial values of these variables determines the initial distribution of facts between nodes.[1] In the experiments reported below (which used the binary tree example introduced in the previous section, all derived (non-leaf) variables were initialised to *false*, and only the allocation of leaves to each node was varied.

The actions of firing a rule, copying a fact from another node and idling are encoded as a Mocha *atom* which describes the initial condition and transition relation for a group

---

[1] We can also leave the initial allocation of facts undetermined, and allow the model checker to find an allocation which satisfies some property, e.g., that there is a derivation which takes less than k steps. However for the experiments reported here, we specified the initial assignment of facts to nodes.

of related state variables. Inference is implemented by marking the consequent of a rule as present in working memory at the next cycle if all of the antecedents of the rule are present in working memory at the current cycle. A rule is only enabled if its consequent is not already present in working memory at the current cycle. Communication is implemented by copying the value representing the presence of a fact in the working memory of another node at the current cycle to the corresponding state variable in the node performing the copy at the next cycle and incrementing a counter, $n_i\_counter$, for the node performing the copy. Copying is only enabled if the fact to be copied is not already in the working memory of the node performing the copy. In the experiments, we assumed that all rules are believed by all nodes in the initial state, and did not implement copying rules. However, this can be done in a straightforward way by adding an extra boolean variable to the premises of each rule, and implementing copying a rule as copying this variable. To express the communication bound, we use a counter for each node which is incremented each time a copy action is performed by the node. To allow a node to idle at any cycle, the atoms which update working memory in each node are declared to be *lazy*.

Mocha supports hierarchical modelling through composition of *modules*. A module is a collection of atoms and a specification of which of the state variables updated by those atoms are visible from outside the module. In our encoding, each node is represented by a module. A particular distributed rule-based system is then simply a parallel composition of the appropriate node modules.

The evolution of the system's state is described by an initial round followed by an infinite sequence of update rounds. The variables are initialised to their initial values in the initial round and new values are assigned to the variables in the subsequent update rounds. At each update round, Mocha non-deterministically chooses between the enabled rules and copy operations, and idling for each node.

The specification language of Mocha is ATL. We can express properties such as 'node i may derive fact ϕ in k steps' as $EX^k\alpha$, where EX^k is EX repeated k times, and α is a state variable encoding of the fact that ϕ is present in node i's working memory (e.g. : $\alpha = n_iA_j$ if $\phi = A_j$).[2] To bound the number of messages used, we can include a bound on the value of the message counter of one or more nodes in the property to be verified. For example, the property 'node i may derive fact ϕ in k steps using at most one message' can be encoded as $EX^k(\alpha \wedge c)$ where c is a boolean variable which is true if $n_i\_counter < 2$. To obtain the actual derivation, we can verify an invariant which states that α is never true, and use the counterexample trace generated by the model-checker to show how the system reaches the state where α is proved.

4 Experimental Results

In this section we describe the results of experiments for different sizes of the binary tree example and different distributions of leaves between the nodes. The experiments

---

[2] In [3] we showed that, given a distributed reasoning system with m nodes, p propositional variables, r propositional rules, and t the largest upper bound on the inference transition in any node, the problem of whether such a temporal property ϕ is true in the system is decidable in time $O(|\phi| \times 2^{m(p+r)} \times t^m)$.

Table 1. Resource requirements for optimal derivation in 8 leaves cases

Case	Node 1	Node 2	# steps	# messages node 1	# messages node 2
1.	A_1-A_8		7	–	–
2.	A_1-A_7	A_8	6	0	3
3.	A_1-A_7	A_8	6	1	2
4.	A_1-A_7	A_8	7	1	1
5.	A_1-A_7	A_8	8	1	0
6.	A_1-A_6	A_7,A_8	6	0	2
7.	A_1-A_6	A_7,A_8	6	1	1
8.	A_1-A_6	A_7,A_8	7	1	0
9.	A_1-A_4	A_5-A_8	5	1	0
10.	A_1,A_3,A_5,A_7	A_2,A_4,A_6,A_8	7	2	3
11.	A_1,A_3,A_5,A_7	A_2,A_4,A_6,A_8	11	0	4

were designed to investigate trade-offs between the number of steps and the number of messages exchanged (a shorter derivation with more messages or a longer derivation with fewer messages).

Table 1 shows the number of derivation steps and the number of messages for each node for varying distributions of 8 leaves. Note that there are several optimal (non-dominated) derivations for the same initial distribution of leaves between the nodes. For example, when node 1 has all the leaves apart from A_8, and node 2 has A_8, the obvious solution is case 5, which requires 1 message and 8 time units: node 1 copies A_8 from node 2, and then derives the goal in 7 inference steps. However, the nodes can solve the problem in fewer steps by exchanging more messages. For example, case 2 describes the situation when node 2 copies A_7 from node 1, while node 1 derives B_3 (step 1). Then node 2 derives B_4 while node 1 derives B_2 (step 2). Then node 2 copies B_3 from node 1, while node 1 derives B_1 (step 3). At the next step node 1 derives C_1 and node 2 derives C_2 (step 4). Then node 2 copies C_1 from node 1 (step 5) and node 1 idles; finally at step 6 node 2 derives D_1. This derivation requires 6 time steps and 3 messages. The trade-off between steps and messages varies with the distribution, as can be seen in cases 10 and 11: if node 1 has all the odd leaves and node 2 all the even leaves, then to derive the goal either requires 7 steps and 5 messages, or 11 steps and 4 messages.

Similar trade-offs are apparent for a problem with 16 leaves, as shown in Table 2. However in this case there are a larger number of possible distributions of leaves, and, in general, more trade-offs for each distribution. The trade-offs are also more dramatic, for example in the 'odd and even' case (cases 20 and 21), where node 1 has all the odd leaves and node 2 all the even leaves, increasing the message bound by 1 reduces the length of the derivation by 10 steps.

Although these examples are very simple, they point to the possibility of complex trade-offs between time and communication bounds in distributed rule-based systems. For more complex examples, we would anticipate that such trade-offs would be harder to predict *a priori*, and our framework would be of correspondingly greater utility.

Table 2. Resource requirements for optimal derivation in 16 leaves cases

Case	Node 1	Node 2	# steps	# copy 1	# copy 2
1.	A_1-A_{16}		15	–	–
2.	A_1-A_{15}	A_{16}	12	0	6
3.	A_1-A_{15}	A_{16}	12	1	4
4.	A_1-A_{15}	A_{16}	13	1	3
5.	A_1-A_{15}	A_{16}	14	1	2
6.	A_1-A_{15}	A_{16}	15	1	1
7.	A_1-A_{15}	A_{16}	16	1	0
8.	A_1-A_{14}	A_{15},A_{16}	11	0	5
9.	A_1-A_{14}	A_{15},A_{16}	11	1	4
10.	A_1-A_{14}	A_{15},A_{16}	12	1	3
11.	A_1-A_{14}	A_{15},A_{16}	13	1	2
12.	A_1-A_{14}	A_{15},A_{16}	14	1	1
13.	A_1-A_{14}	A_{15},A_{16}	15	1	0
14.	A_1-A_{12}	$A_{13},A_{14},A_{15},A_{16}$	11	0	4
15.	A_1-A_{12}	$A_{13},A_{14},A_{15},A_{16}$	11	1	2
16.	A_1-A_{12}	$A_{13},A_{14},A_{15},A_{16}$	12	1	1
17.	A_1-A_{12}	$A_{13},A_{14},A_{15},A_{16}$	13	1	0
18.	$A_1-A_3,A_5-A_7,A_9-A_{11},A_{13}-A_{15}$	A_4,A_8,A_{12},A_{16}	13	2	6
19.	$A_1-A_3,A_5-A_7,A_9-A_{11},A_{13}-A_{15}$	A_4,A_8,A_{12},A_{16}	19	4	0
20.	$A_1,A_3,A_5,A_7,A_9,A_{11},A_{13},A_{15}$	$A_2,A_4,A_6,A_8,A_{12},A_{14},A_{16}$	13	4	5
21.	$A_1,A_3,A_5,A_7,A_9,A_{11},A_{13},A_{15}$	$A_2,A_4,A_6,A_8,A_{12},A_{14},A_{16}$	23	0	8

5   Related Work

The upper limit on deliberation (or response) time in rule-based systems is a well-established problem. However previous work has studied expert and diagnostic systems as single isolated systems [4], and has focused mainly on termination (or worst case response time), rather than more general issues of resource bounds, and trade-offs between time and communication resources.

There exists considerable work on the execution properties of rule based systems, both in AI and in the active database community. In AI, perhaps the most relevant work on the execution properties of rule based systems is that of Cheng and co-workers on predicting the response time of OPS5-style production systems. For example, in [5], Chen and Cheng show how to compute the response time of a rule-based program in terms of the maximum number of rule firings and the maximum number of basic comparisons made by the Rete network. In [6], Cheng and Tsai describe a tool for detecting the worst-case response time of an OPS5 program by generating inputs which are guaranteed to force the system into worst-case behaviour, and timing the program with those inputs. However the results obtained using these approaches are specific to a particular rule-based system (OPS5 in this case) and cannot easily be extended to systems with different rule formats or rule execution strategies. Nor are they capable of dealing with the asynchronous inputs found in communicating RBSs.

Another relevant strand of work is the problem of termination and query boundedness in deductive databases [7]. However, again this work considers a special (and rather restricted with respect to rule format and execution strategy) class of rule-based systems. In our previous work, e.g., [8,9] we have investigated time vs. memory trade-offs for single (rule-based and resolution) reasoners, and in [10], we investigated resource requirements for time, memory and communication for systems of distributed resolution reasoners. In [3] we showed how durations can be assigned to the various stages of the inference cycle (matching, conflict resolution etc.) and how abstraction techniques can be used to model sets of individual rule firings into a single abstract transition with associated upper and lower time bounds.

6   Conclusions

In this paper, we proposed an approach to modelling and verifying resource requirements of distributed rule-based systems. We described results of experiments on a synthetic example which show interesting trade-offs between time required by the nodes in a distributed rule-based system to solve the problem and the number of messages they need to exchange. The paper presents initial results of a long-term research programme. In future work, we plan to evaluate our approach on real-life examples of rule-based systems.

References

1. Alur, R., Henzinger, T.A., Mang, F.Y.C., Qadeer, S., Rajamani, S.K., Tasiran, S.: MOCHA: Modularity in model checking. In: Y. Vardi, M. (ed.) CAV 1998. LNCS, vol. 1427, pp. 521–525. Springer, Heidelberg (1998)
2. Clarke, E.M., Grumberg, O., Peled, D.A.: Model Checking. MIT Press, Cambridge (1999)
3. Alechina, N., Logan, B.: Verifying bounds on deliberation time in multi-agent systems. In: Proceedings of the Third European Workshop on Multiagent Systems (EUMAS 2005), pp. 25–34 (2005)
4. Georgeff, M.P., Lansky, A.L.: Reactive reasoning and planning. In: Proceedings of the Sixth National Conference on Artificial Intelligence, AAAI 1987, pp. 677–682 (1987)
5. Chen, J.R., Cheng, A.M.K.: Predicting the response time of OPS5-style production systems. In: Proceedings of the 11th Conference on Artificial Intelligence for Applications, p. 203 (1995)
6. Cheng, A.M.K., yen Tsai, H.: A graph-based approach for timing analysis and refinement of OPS5 knowledge-based systems. IEEE Transactions on Knowledge and Data Engineering 16, 271–288 (2004)
7. Brodsky, A., Sagiv, Y.: On termination of datalog programs. In: International Conference on Deductive and Object-Oriented Databases (DOOD), pp. 47–64 (1989)
8. Albore, A., Alechina, N., Bertoli, P., Ghidini, C., Logan, B., Serafini, L.: Model-checking memory requirements of resource-bounded reasoners. In: Proceedings of the Twenty-First National Conference on Artificial Intelligence (AAAI 2006), pp. 213–218 (2006)
9. Alechina, N., Bertoli, P., Ghidini, C., Jago, M., Logan, B., Serafini, L.: Verifying space and time requirements for resource-bounded agents. In: Proceedings of the Fourth Workshop on Model Checking and Artificial Intelligence (MoChArt 2006), pp. 16–30 (2006)
10. Alechina, N., Logan, B., Nga, N.H., Rakib, A.: Verifying time, memory and communication bounds in systems of reasoning agents. In: Proceedings of the Seventh International Conference on Autonomous Agents and Multiagent Systems (AAMAS 2008), pp. 736–743 (2008)

Meta-analysis for Validation and Strategic Planning

David A. Ostrowski

System Analytics and Environmental Sciences
Research and Advanced Engineering
Ford Motor Company
dostrows@ford.com

Abstract. This paper presents framework to support design of meta-rule constructs. A prototype is described towards the application of credit analysis. The focus of this system is to define a higher level of inference that will guide preestablished object-level rule constructs. This architecture is supported by the incorporation of machine-learning (ML) techniques to support the acquisition of business rule-based knowledge. This knowledge is applied to specific parameters that ultimately guide an object level decision making process. By developing this process of automated knowledge acquisition, we are interested in up front actions including validation and support of intended responses. We also intend to further classify and categorize this acquired knowledge to support future policy modifications.

Keywords: metaknowledge, metaheuristics, expert systems, strategic knowledge, common sense knowledge.

1   Introduction

Metaknowledge is knowledge about the control and use of domain knowledge in an expert system. In its simplest form, metaknowledge can be defined as knowledge about knowledge [1]. Metaknowledge can be defined within a generalized framework consisting of three categories: Doing, Reasoning and Metareasoning (figure 1.). In this common description, reasoning is defined as controlling action at the ground level where the meta-level applies control to where reasoning is sufficient and deciding when action can proceed.

One of the earliest applications of meta-level understanding is the MYCIN application [3]. Established as a tool for expert analysis in medical infections, the MYCIN application applied a higher level of control through a specific application of properties to how a rule accomplishes a particular goal. Davis developed these techniques further, by supporting a higher level of distinction between rules [4]. He described a three step process towards the development of a meta-level control strategy including retrieval, refinement and execution. Within the retrieval phase, patterns are used to select a knowledge source (rules). During refinement, the set of knowledge rules are re-ordered to support a finer degree of control. In the final phase, one can reserve a knowledge source in the revised rule set to be applied to the problem. Expanding on this work, Mussi developed a method for the representation of common sense in

N. Bassiliades, G. Governatori, and A. Paschke (Eds.): RuleML 2008, LNCS 5321, pp. 39–46, 2008.
© Springer-Verlag Berlin Heidelberg 2008

diagnostic expert systems. His work extended the rule-based approach to the preferences and maintenance of human expert beliefs applying, often unconsciously, heuristic considerations. The application of this belief mechanism was referred to as 'strategic common sense' [5][6].

The PARULEL system supported parallel execution of rule languages that could be used to effectively determine the allocation of mortgage pools. This technique was found to be highly advantageous in dealing with this combinatorial optimization problem as applied by financial institutions that trade in mortgage-backed securities [7]. Metaheuristics have also found utility in simulation based environments. Work by Hoogendoorn in this area clearly expanded earlier interpretation towards a framework defined as objects, meta and meta-meta rules [8].

We are interested in development of meta-information in support of a plan revision and recognition as well as the area of development of knowledge-sharing strategies. Our interest includes the inclusion of all relevant information concerning the guiding of the current 'common-sense' strategies. Towards these learning mechanisms, we have two specific goals. The first goal consists of identifying or predicting suitable ranges as values for our control strategy. We view the capability of predicting specific parameters and their ranges to be useful for initially validating information and at a secondary level, provide assistance towards the guiding of object-level heuristics. The second goal is to take our established prediction model and apply to the categorization of our existing data to support the future policy making.

We view the distinctive metaknowledge rules as a clearly defined AI problem. We examine the activity of rule derivation as a separate search problem. Drawing inspiration from hybrid systems we see the development of control parameters as a unique opportunity to the future expansion of expert systems analysis [9]. In Section Two we present our description of a generalized framework. Section Three explains the development and utilization of our prototype system and in section four we present our conclusions.

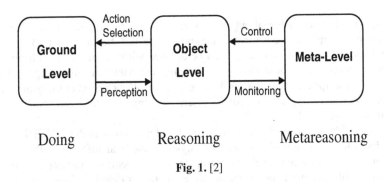

Fig. 1. [2]

1.1 AI and Strategic Knowledge

Metareasoning environments can be viewed as the construction of small worlds that attempt to approximate larger world models. In this construction, it is assumed that the underlying state space is known and well defined. When reasoning within such a model Laskey et. al. addressed four issues. 1) formulating the problem 2) deciding whether to trust the results 3) knowing what to do when you have decided not to trust your model 4) building a better small world. In addressing these concepts, it is ruled that decision theory does not currently treat any of these metareasoning issues [10]. This is consistent with studies involving computational decision models which find that people do not manage uncertainty in ways which closely resemble probalistic reasoning [11]. It is within these constructs that we find the higher opportunity for further application of learning methods at the metareasoning level. The reasoning within such a model can be identified as the study of perturbation tolerance – or the ability of a system to quickly recover after an unexpected event. This can also be described as 'brittleness' [12].

Prior knowledge has also been related towards the application of guiding learning mechanisms in order to more effectively deal with real world situations. Multitask learning has incorporated dependencies between tasks into an inductive learning system [13]. A multitask learning system uses the input-output examples of related tasks in order to improve the general performance. This approach has demonstrated value in terms of being able to overcome characterizations of non-stationarity and extreme noisiness. This is demonstrated in the ability to define case-based reasoning at the object-level means of implementation. This level of introspection has provided for an effective means of determining the data [14].

The maintenance of prior knowledge supports the concept of beliefs which have long been applied evolutionary computation solutions. Reynolds introduced a framework in the context of evolutionary computation in which beliefs were maintained and influenced by each generation [15]. Termed as Cultural Algorithms, they modeled the characteristic of a belief system or knowledge that is maintained over a biologically inspired evolutionary process. Representations of this environment included applications of hierarchical structures for the purpose of maintaining abstractions of data [16].

We feel that the most successful approach towards development of a framework that can suit our needs is to be able to develop a mechanism that supports a self-guided approach to the development of a perturbation-tolerant learning mechanism. This is accomplished through integrating a learning process to support the development of control parameters.

2   A Generalized Framework

In figure 2, we present a generalized framework which we use to design our prototype. The components denote processes, solid lines denote information and dotted lines denote a separation between the hierarchy of information. The bottom level (application) describes interaction with the real world which includes the end user application, relating data from the real world as well as feedback from the interaction with the domain experts. The middle level is the execution of our heuristical information implemented as rules to provide decision making support to the application layer. This level represents implementation of a traditional expert-system knowledge base. The upper middle level identifies the 'control' rules. This distinct layer can be considered as rules that define the KB rules. A self guided learning mechanism adjusts parameters to be used by the control rules influenced by a feedback loop. The control rules are influenced by a (ML-based) model derived from historical data and maintained (retrained) by current observations. The output from the model support prediction and validation for static parameters as well as ranges to be applied in the context of fuzzy logic. The highest level of strategic heuristics identifies a set of rules from which to make a selection between learning components. A secondary planning component involves the selection of learning components to be applied to gain convergence. In this high level design, we support the distinction of levels, especially between the meta-levels and the regular knowledge-based levels. By making a clear separation between these levels, this component-based architecture allows parallel comparisons between learning components.

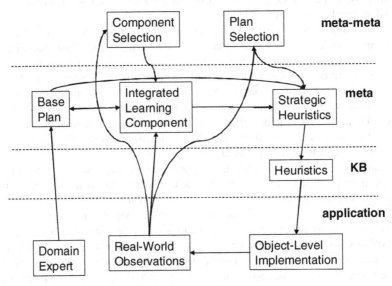

Fig. 2. Metarules framework

3 Prototype Description

The focus of our prototype system is an implementation of the (third) meta-level as presented in our generic framework. This layer interacts with the expert system that supports the credit approval application (representing an implementation of the bottom two layers). Our prototype description begins with a process level description of the application environment. The first step of this process begins when a customer initiates a credit application via a dealer or internet. The credit application process references various databases in order to consolidate information from credit agencies, trade-in data and customer history for both the applicant and co-applicant. The credit application process produces relevant information including an internally designed scoring mechanism that is compared with credit bureau scoring along with terms of the credit contract. This system is assisted by a backend knowledge base system that applies known expert level heuristics to the same data set. Through the application of fuzzy logic and case based reasoning this rule based system is utilized to detail specific strengths and weaknesses of a deal. Here, expert-level analysis can be provided to ensure a higher level of quality and consistency across all credit applications.

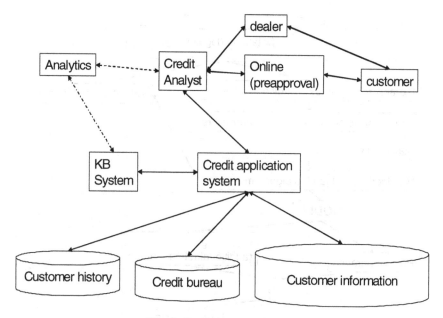

Fig. 3. Credit Approval Process

Our system accomplishes this task by applying machine learning techniques against historical data for the purpose of deriving improved individual parameters to be applied to rules as well as parameter ranges to be applied in fuzzy rule constructs. By establishing predictive parameters towards the output, we are able to formulate 'control' based rules to be applied to the expert system heuristics. As a result, indicated output to an analyst can range from warnings predicated on the likelihood of data inconsistency as well as the improved means of strategies that are suggested. Communication paths of this activity are indicated by the dotted lines on figure 3.

3.1 Detail of System Design

The process of developing our model is presented in figure 4. The first step is to extract necessary parameters from a complete credit deal. Here, all relevant parameters (bureau, credit terms, trade-in, rates) are derived from the database either through the means of rule-based inference or generation from composite rules. The parameters undergo a level of verification against initial data transaction sets. After the data is verified, a machine learning algorithm is trained using a set of historical data. Once our model is established within an appropriate range, we can then apply it to the construction of basic rules. Initial rule generations derived from this effort include the comparisons between a predictive value along with the values obtained from the credit application process. The parameters are applied in terms of controlling object rules through the reinforcement of initial strengths and weaknesses of a transaction. If a parameter is not appropriate it may reinforce an established hypothesis therefore helping the analyst quickly obtains a more comprehensive view of a deal.

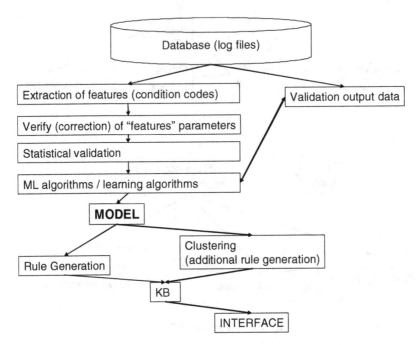

Fig. 4. Process of Model Generation for Strategic Heuristic Rules

3.2 Examples

3.2.1 Validation of Income

One example is determining a model to verify stated income on a consumer credit application. In this particular example, a set of parameters is established that correlates to the levels of debt, prior credit transactions, credit scoring and other credit history (such as mortgages) to determine the support towards this predictive variable. Here, we applied a nearest neighbor algorithm with regards to estimating stated income. A corresponding rule is then established generating a warning sign to the analyst. Next the information is applied towards cluster-based (k-means) analysis for the purpose of determining significant characterizations among the proposed credit contract in order to further identify specific classifications that determine our predictive variables.

3.2.2 Response Model

In this example, we examine the requirements with regards to a credit offering structure by examining competitive information (rates), demographical information (users) and terms through the application of a neural net approach. This focus evaluations our credit policy towards structure of a deal in order to best support a successful return. This is accomplished by deriving a predictive model to the likelihood of purchasing of future contracts. Through identifying our model, an analyst will receive a indicator that an offer (based on past history data) that his deal structure (terms) is not likely to result in a purchased (or not purchased) contract.

4 Conclusion

A complete framework for a meta-knowledge architecture has been presented. A prototype implementation of the initial meta-layer has been described. Our prototype is currently being tested in production and applied towards the validation of parameters. Among the ML algorithms that have been applied and compared among each other is the nearest-neighbor, neural-net (backpropagation) and genetic programming. Work also includes the application of clustering approaches and support vector machines to classify the data in order to further structure the credit contact. Considerations are also being made towards the visualization of the data in order to assist in the continued interaction of the analyst in the acquisition of expert knowledge.

Future work includes the incorporation of expanded data sources through the inclusion of web-generated demographic information to further develop consumer profiles. Current experiments have demonstrated an improved level of support towards further diagnosis of credit transactions. Improved support lends itself to applications of the semantic web, where additional consumer, demographic and competitive information is provided by the utilization of a web based format.

References

1. Waterman, D.A.: A Guide to Expert Systems. Addison-Wesley, Reading (1986)
2. Cox, M.T., Raja, A.: Metareasoning: A Manifesto, Technical Report, BBN TM-2028 BBN Technologies (2007)
3. Adams, J.B.: A Probability Model of Medical Reasoning and the MYCIN model, Math. Biosciences 32, 177–186 (1976)
4. Davis, R.: Meta-Rules: Reasoning about Control. Artificial Intelligence, 179–220 (1980)
5. Mussi, Silvano, Morpurgo, Rosamaria: Acquiring and Representing knowledge in the diagnosis domain. Expert Systems 7(3) (August 1990)
6. Mussi, Silvano: A Method for Putting Strategic Common Sense into Expert Systems. IEEE Transactions On Knowledge and Data Engineering 5(3) (June 1993)
7. Stolfo Salvadore J., Chan Philip K., Woodbury Leland, G., Ohsie, D.: PARULEL: Parallel rule processing using metarules for redaction. Journal of Parallel and Distributed Computing (1991)
8. Hoogendoorn, M., Jonker, C.: A Meta-Level Architecture for Strategic Reasoning in Naval Planning. In: Ali, M., Esposito, F. (eds.) IEA/AIE 2005. LNCS (LNAI), vol. 3533, pp. 848–850. Springer, Heidelberg (2005)
9. Negnevitsky, Michael: Artificial Intelligence A Guide to Intelligent Systems. Addison Wesley, Reading
10. Laskey, Blackmond, K., Lehner, Paul, E., Lehner: MetaReasoning and the Problem of Small Worlds. IEEE Transactions on Systems, Man and Cybernetics 24(11) (November 1994)
11. Anderson Michael, L., Perlis Donald, R.: Logic, self-awareness an self-improvement. The metacognitive loop and the problem of brittleness. Journal of Logic and Computation 14 (2004)
12. Fox, J.: Making Decisions Under Inuence of Knowledge, Psychological Review, Barthai, Kai, Gutjahr, Steffen, Nakhaeizadeh, Gholamena, Incorporating Prior Knowledge About Financial Markets Through Neural Multitask Learning, pp. 191–121

13. Bartlmae, K., Gutjahr, S., Nakhaeizadeh, G.: Incorporating prior knowledge about financial markets through neural multitask learning. In: Proceedings of the Fifth International Conference on Neural Networks in the Capital Markets (1997)
14. Bartlmae, K., Gutjahr, S., Nakhaeizadeh, G.: Incorporating prior knowledge about financial markets through neural multitask learning. In: Proceedings of the Fifth International Conference on Neural Networks in the Capital Markets (1997),
    http://citeseer.ist.psu.edu/bartlmae97incorporating.html
15. Reynolds, R.G.: An Introduction to Cultural Algorithms. In: Fogen, A.V. (ed.) The Proceedings of the 3rd Annual Conference on Evolutionary Programming, L.H. River Edge, NJ, pp. 131–139. World Scientific Publishing, Singapore (1994)
16. Isles, R.: Reasoning About Genetic Program Function and Structure Using Cultural Algorithms, Masters Thesis, Wayne State University (2000)

Abductive Workflow Mining Using Binary Resolution on Task Successor Rules

Scott Buffett

National Research Council Canada, Fredericton, NB, E3B 9W4
Scott.Buffett@nrc.gc.ca

Abstract. The notion of *abductive workflow mining* is introduced, which refers to the process of discovering important workflows from event logs that are believed to cause or explain certain behaviour. The approach is based on the notion of abductive reasoning, where hypotheses are found that, if added to a rule base, would necessarily cause an observation to be true. We focus on the instance of workflow mining where there are critical tasks in the underlying process that, if observed, must be scrutinized more diligently to ensure that they are sufficiently motivated and executed under acceptable circumstances. Abductive workflow mining is then the process of determining activity that would necessarily imply that the critical activity should take place. Whenever critical activity is observed, one can then inspect the abductive workflow to ascertain whether there was sufficient reason for the critical activity to occur. To determine such workflows, we mine recorded log activity for task successor rules, which indicate which tasks succeed other tasks in the underlying process. Binary resolution is then applied to find the abductive explanations for a given activity. Preliminary experiments show that relatively small and concise abductive workflow models can be constructed, in comparison with constructing a complete model for the entire log.

1 Introduction

The diverse sets of advantages and disadvantages that various existing workflow mining algorithms have to offer demonstrate that there is seemingly a solution to fit almost any need. Techniques such as van der Aalst's α-algorithm [8], $\alpha++$ algorithm [11] and heuristic miner [10] offer mining techniques that produce concise, simplified workflows that are easy to visualize, but sacrifice accuracy. Still others such as Petrify [7] and Region Miner [9] achieve high fitness and precision, but as a result can produce workflows that are too large or complicated to be reasonably understood by a human viewer. In this case, reducing the size and complexity of a workflow model by eliminating components that are not of interest to a particular observer with a specific purpose for viewing the model, while retaining the desirable qualities of fitness and precision, can be quite advantageous. Accomplishing this is the focus of this paper.

Situations where an observer needs to view and understand very specific components of a workflow model for a particular purpose are prevalent in the domain

N. Bassiliades, G. Governatori, and A. Paschke (Eds.): RuleML 2008, LNCS 5321, pp. 47–57, 2008.

of compliance. In this case, the primary focus commonly seen in workflow mining of understanding how processes work so that they can be improved upon, made more efficient, or simplified, is not as crucial. Instead, the goal here lies in ensuring that individuals' behaviour is deemed appropriate within the context of the overall observed activity. Thus the workflow only needs to be analyzed to the extent that it can be determined whether a particular event is compliant with the rules, given the observed activity within which the event is situated. This greatly simplifies the compliance checking process. It is then not necessary to collect all tasks in a particular case and check against the entire workflow model to ensure consistency. Instead only the simplified model, which shows the legal traces of activity with which a particular action may be associated, needs to be analyzed. Using the simplified model will improve accuracy by reducing false positive and false negative cases, since reducing the model should naturally reduce the number of structures in the model that degrade fitness and precision, such as loops. Just as important, this reduction will also greatly simplify the cognitive burden imposed on the viewer of the workflow model. If inappropriate activity is detected, the graphical representation of the simplified model can be displayed to a compliance officer, who will more easily be able to understand why a particular action was deemed inappropriate, and quickly make a decision on the subsequent course of action.

In this regard, the task is not as simple as merely isolating a specified region or subgraph of the workflow model. Given a particular activity for which compliance checking is desirable, it needs to be determined exactly what part of the overall workflow model is needed. To accomplish this task, we introduce the concept of *abductive workflow mining*. This concept comes from the particular field of logical reasoning known as *abductive reasoning*. Abductive reasoning refers to the problem of of finding *explanations* for a particular set of facts. These explanations are found to be sets of facts and axioms that, if true, will logically imply that the original set of facts are true. We apply this concept to workflow mining by discovering workflows that, if followed in a particular trace, would offer an explanation for why a particular task was executed. Such a workflow is known as an *abductive workflow*.

In this paper, we present a method that finds abductive workflows by mining and constructing propositional task successor rules. These rules, which are represented in Kowalski form, indicate which tasks follow which other tasks in the underlying process that is being mined. Given a simple action or a complex set of activity, binary resolution is applied to the rules to determine abductive explanations for the activity in question. The set of explanations can then be simplified to form a minimal set. We choose the stronger act of identifying abductive explanations, rather than simply identifying consistent activity, due to the possibly critical nature of the activity in question. For example, if the critical activity was the act of a bank employee accessing sensitive customer data, such as a credit report, the idea is to identify activity that would necessarily imply that the critical activity would have to take place (such as a loan application being processed), rather than to identify activity that is merely consistent with the activity in question (such as the credit report being printed). Compliance software or a privacy officer could then do a simple check to ensure that at least one such causal activity (or sequence of activities, i.e. workflow) was indeed executed.

The remainder of the paper is presented as follows. In section 2 we discuss required background theory on workflow mining and abductive reasoning that we employ in our techniques. Section 3 then defines abductive workflow, and introduces some necessary theory surrounding the concept. In section 4 we outline our logic-based technique for discovering abductive workflows, and in section 5 we presents a few results demonstrating the performance of our method. Section 6 then discusses conclusions and offers a few directions for future work.

2   Background

2.1   Workflow Mining

Workflow mining [2] refers to the process of autonomously examining a transaction log of system events, and extracting a model of the underlying process being executed by the events. Generally speaking, an log consists of a number of events, each of which being associated with a *task* and a *case*. An event's task refers to the actual activity the event represents, while the event's case refers to the instance of the underlying business process to which the event belongs. Each case in the log consists of a sequence of tasks (often referred to as a *trace*) that represent a complete and ordered set of actions that are executed in an instance of the business process. Workflow mining techniques are then used to build a model of the business process by representing the different ways a case in the process can be executed. A number of different representations have been used in the literature, perhaps the most common of which being the Petri net [3,1]. Figure 1 represents a small example log, as well as the resulting Petri net representing the mined workflow. Any legal sequence of transitions that takes a token from the start (leftmost) place to the end (rightmost) place represents a different way of executing the business process. Thus by a quick inspection of the graphical model, one can infer different characteristics of the process, such as the fact that C and D can be executed interchangeably, due to the preceding "OR-split" where one place points to the two transactions, or that both B and either C or D can be executed in parallel after A, due to the preceding "AND-split" where A points to two places.

2.2   Abductive Reasoning

Abduction is an instance of assumption-based reasoning where the focus is to determine a hypothesis (or a set of hypotheses) that, if true, would necessarily explain an observation or evidence in question [4]. This type of reasoning is particularly useful when one observes an interesting or curious event or happening,

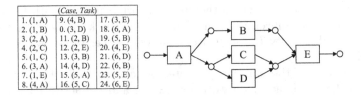

(Case, Task)		
1. (1, A)	9. (4, B)	17. (3, E)
2. (1, B)	0. (3, D)	18. (6, A)
3. (2, A)	11. (2, B)	19. (5, B)
4. (2, C)	12. (2, E)	20. (4, E)
5. (1, C)	13. (3, B)	21. (6, D)
6. (3, A)	14. (4, D)	22. (6, B)
7. (1, E)	15. (5, A)	23. (5, E)
8. (4, A)	16. (5, C)	24. (6, E)

Fig. 1. Example log and corresponding workflow diagram

and may want to know what could possibly cause the event to be true. That is, given evidence E and rule base R, abductive reasoning attempts to find facts that, when added to R, would necessarily imply E. Such facts are then likely to have a causal effect on E in the real-world application being modeled by the rule base.

Abductive reasoning is of particular importance in fault diagnosis in complex systems. Here the rule base represents the operational characteristics of the system, and an observation is typically some fault or error in system function. Abductive reasoning is then used to determine what sort of activities would necessarily cause the fault to occur. These activities are then considered to be the primary causes of the error, and are thus investigated first. So abduction is applied to diagnosis with the idea that actions that would necessarily cause a fault to occur are more likely to be the true cause than actions that are merely consistent with the fault occurring.

We consider the application of abductive reasoning to workflow analysis in a similar way. Here the complex system to be modeled is the business process being analyzed, and the activities in question are the more serious or potentially harmful tasks that are executed within the process. Such critical tasks should not be executed freely; rather there should be an impelling reason for such tasks to be performed. Abductive reasoning can then be applied to determine what sort of activities would be sufficient for validating the critical behaviour in question, allowing one to easily verify whether the behaviour was warranted in any particular situation.

3   Abductive Workflow

Let W be a workflow model constructed based on a set T of tasks and a set C of cases of executions of T, as defined in section 2.1 above. Additionally, let W' be a sub-workflow of W, for which an explanation is desired. Note that W' may often simply consist of a single transition, representing the event to be explained, but in general could be any workflow structure. We consider a workflow W_a to be an *abductive workflow* of W for W' if and only if any activity observed in W_a necessarily implies that there is associated activity in W'. This means that observing activity in W_a necessarily implies that activity in W' will occur, and thus W_a is an explanation for W'.

Note that one might think that an easy way to find an abductive workflow that satisfies this condition is to simply set $W_a = W'$. So, for example, one could say that an explanation for a task A be executed is to observe that A is executed. Clearly, something stronger is needed. We need to find what sort of extra activity, not including A, would cause A to be executed. So, generally speaking, an explanation for W' is not allowed to include W'. W' may reside within the abductive model, but the activity surrounding W' must be sufficient for W' to occur. In this section we lay out formal definitions for the properties such as this for a workflow model to be considered an abductive workflow. In the following section we then offer general ideas for how an abductive workflow model can be constructed.

Definition 1 (consistent). *Let C be a set of cases and let W be a workflow model (not necessarily built based on C). A case $c \in C$ is said to be consistent with W if there is a valid trace through W that is a (non-empty) substring of c. Moreover, any sequence t of events is consistent with W if there is a valid trace through W that is a (non-empty) substring of t.*

So a case c is consistent with W' if observing c would necessarily cause a trace through W' to be observed, perhaps with extra activity from c occurring before and/or after W'. Let $C^+ \subseteq C$ then be the set of cases in C that are consistent with W', and accordingly let $C \setminus C^+$ be denoted by C^-. Any case in C^+ being observed will then necessarily cause activity in W' to occur. We use this partition over the set as the basis for the abductive model. First we define an *abductive trace*, which is the key concept in modeling abductive workflow.

Definition 2 (abductive trace). *Let W' be a workflow model and let C be a set of cases partitioned into C^+ and C^- as above. A sequence t_a of events is an abductive trace in C for W' if and only if each of the following hold:*

- *t_a is a substring of some $c^+ \in C^+$*
- *For any $c^- \in C^-$, (1) t_a is not a substring of c^-, and (2) if any possible trace $t_{W'}$ through W' is a substring of t_a, then there is no string S such that the result of replacing $t_{W'}$ with S makes t_a a substring of c^-.*

So, not only does an abductive trace t_a cause activity in W', the part of t_a without the activity from W' necessarily causes activity in W'. This means that the activity in t_a that resides in W' cannot be replaced, and thus the activity in W' *must* occur. To illustrate the necessity for this restriction, consider a simple example with two cases $ABDE$ and $ACDF$, with W' simply being B. Then $ABDE$ is a positive case and $ACDF$ is a negative case. The trace ABD would not be considered an abductive trace, since executing B was not necessary in this situation. Another choice was possible, namely D. Put another way, for any event x, if A is performed immediately before x and D is performed immediately after, it is not necessarily the case that x would be B. However, BDE on the other hand would be an abductive trace (as would $ABDE$, DE, or even simply E), since it is necessarily the case that an event x followed immediately by DE would have to be B. Using the notion of abductive traces, we define abductive workflow:

Definition 3 (abductive workflow). *Let C be a set of cases yielding workflow W. A workflow W_a is an abductive workflow of W for workflow W' if and only if every valid trace through W_a is an abductive trace in C for W'.*

Thus any activity that is consistent with W_a, minus any activity that forms a valid trace through W', will necessarily cause a valid trace through W' to be executed.

   While any abductive workflow will specify activity that will explain the activity in question, there may be additional explanations that are not captured by the model. That is, it is possible that activity in W' may occur without any explanation from W_a. Ideally, we would prefer an abductive workflow that would capture explanations for every case. That is not to say that we would expect W_a to capture every possible explanation; rather, it would be equally useful to generate abductive workflows that, given any activity in W', will contain *some* explanation for that activity. To give an indication of how well an abductive workflow performs at offering such an explanation, we define the completeness measure for an abductive workflow.

Definition 4 (completeness measure). *Let $C = C^+ \cup C^-$ be a set of cases yielding workflow W and let W_a be an abductive workflow of W for W' as defined above. The completeness measure for W_a is the probability that a case in C^+ is consistent with W_a.*

The completeness measure is thus a measure of how likely it is that an occurrence of activity in W' will be explained by W_a, which is equal to the probability that a case containing a trace of activity through W' will also contain a trace of activity through W_a.

Definition 5 (complete). *An abductive workflow with completeness measure equal to 1 is said to be complete.*

Another desirable characteristic of abductive workflow is that the model be small and concise, making it easy to check, as well as less likely to introduce errors. There are two ways to measure the size of a workflow model: (1) by considering the length of the traces through the model, and (2) by considering the number of traces through the model. We consider each of these separately.

Definition 6 (task-minimal abductive trace). *An abductive trace t_a is task-minimal if there is no substring of t_a that is an abductive trace.*

Definition 7 (task-minimal abductive workflow). *An abductive workflow is task-minimal if all abductive traces making up the workflow are task-minimal.*

So a task-minimal abductive workflow allows for only small traces, but may allow for more traces than necessary. We tackle this dimension next.

Definition 8 (trace-minimal abductive workflow). *An abductive workflow is trace-minimal if there is no subgraph of the workflow with the same completeness measure.*

Thus an abductive workflow is minimal if it cannot be reduced in any way, thus reducing the number of allowable traces, without giving up some of the explanations it offers.

Putting it all together, a best possible abductive workflow is one that is:

- complete
- task-minimal
- trace-minimal

and perhaps ideally, has the fewest nodes out of all workflows that meet the above conditions.

4   Discovering Abductive Workflow Using Task Successor Rules

In an effort to find abductive workflows that meet the criteria described in the previous section, we mine the log data for rules, referred to as *task successor rules*, that indicate which activity immediately follows certain tasks in the log. These rules are represented in Kowalski form [6]. A non-horn clause is represented as a rule in Kowalski form if the tail of the rule consists of the conjunction of the atoms of the negative literals of the clause, and the head consists of the disjunction of the positive literals. Thus we construct rules of the form:

$$p_1 \wedge \ldots \wedge p_m \rightarrow q_1 \vee \ldots \vee q_n$$

Such a rule is interpreted as "the set of tasks $\{p_1, \ldots, p_m\}$, when observed as a whole, is always immediately followed by one of the tasks in $\{q_1, \ldots, q_n\}$". So the conjunction of tasks $p_1, \ldots, p_m$ being observed implies the disjunction $q_1, \ldots, q_n$. For convenience, such rules are henceforth denoted as $p_1 \ldots p_m \rightarrow q_1, \ldots, q_n$.

The set of rules is constructed as follows. Let W' represent the activity for which an explanation is desired, and let $C^+ \subseteq C$ be the set of cases consistent with W' and $C^- = C \setminus C^+$, as defined above. Let $C^+_{-W'}$ be a transformation of C^+ where, for every c in C^+, there is a corresponding c' in $C^+_{-W'}$ where c' is equivalent to c with two differences: (1) the specific tasks in c that represent a trace through W' (i.e. the part that makes it consistent with W') are removed, and (2) a dummy task (denoted by "w'"), which replaces the removed trace from W', is appended to the end. For example, let W' be represented by the execution of task A followed by task B. Then the case $KABD$ in C^+ would be represented by KDw' in $C^+_{-W'}$. This then represents that a case with K executed earlier than D, with no other activity (save for activity in W'), can cause activity in W' to occur (since the case includes w'). Also, let $C^-_{W'_o}$ be the cases in C^- with dummy tasks (denoted by "w'_o") appended to the end. Given these transformations, a task successor rule is then created for every subsentence of activity (minus the dummy task w') that appears in a case in $C^+_{-W'}$, with the activities in the subsentence making up the tail of the rule. The head of the rule is then the disjunction of all tasks that immediately follow that set of activity in any case in $C^+_{-W'} \cup C^-_{W'_o}$.

Example 1 (Task Successor Rules). Let the set of cases be $\{PQR, PRS, RMN,$ $TQV\}$, and let W' simply represent the execution of task R. Then $C^+ =$ $\{PQR, PRS, RMN\}$, $C^- = \{TVQ\}$ and thus $C^+_{-W'} = \{PQw', PSw', MNw'\}$ and $C^-_{W'_o} = \{TVQw'_o\}$. The set of task successor rules is then:

$$P \to Q, S \qquad Q \to w', w'_o$$
$$S \to w' \qquad M \to N$$
$$N \to w' \qquad PQ \to w'$$
$$PS \to w' \qquad MN \to w'$$

Once the rules are constructed, they are converted to clauses in conjunctive normal form. Binary resolution is then applied to find new clauses where w' is the only positive literal. Atoms from the negative literals must then necessarily imply w', where w' represents a trace of activity from W'. Thus the tasks represented by these atoms make up an abductive explanation. One could also make use of other mechanisms, such as assumption-based truth maintenance systems, to generate the abductive explanations.

Example 2 (Abductive Traces). Given the task successor rules generated in Example 1, one could perform the following sequence of resolutions and conclude that the presence of task P implies W':

$$\bar{P}QS, \quad \bar{P}\bar{Q}w' \quad \to \quad \bar{P}Sw'$$
$$\bar{P}Sw', \quad \bar{S}w' \quad \to \quad \bar{P}w'$$

The set of abductive traces that would ultimately be found for this example is $\{P, S, M, N, PS, MN, PQ\}$. Thus the observation of any of these seven sequences implies that R will be performed. Note that the inclusion of w'_o in $C^-_{W'_o}$ (and subsequently in the task successor rule $Q \to w', w'_o$) ensures that Q is not chosen as an abductive explanation.

Even from a small example such as this, we see that an unnecessarily large number of explanations can be generated. As mentioned above, it is desirable to obtain abductive traces that are task-minimal. That is, it is more simplified to present either M or N as abductive explanations of R, rather than specify that both M and N are needed. Moreover, it also quite desirable to obtain workflow representations that depict only enough information that will minimally but completely explain W', and are thus complete and trace-minimal. In the above example, both P and S are explanations for R; however, it would be redundant to report both explanations, since P would explain the appearance of R in any case that S would.

Example 3 (Task-Minimal Traces). The set of abductive traces from Example 2 that are task-minimal is $\{P, S, M, N\}$. This is obtained by simply removing the supersets.

Example 4 (Complete Workflows). Using only the task-minimal traces in Example 3, there are six complete abductive workflows. Since each trace in this example consists of only a single event, each abductive workflow representation will consist of a number of possible single-event sequences, and thus are denoted here simply as a list of these events. The complete workflows are PM, PN, PMN, $PSM, PSN, PSMN$. This means that, if event R occurs, each of the six complete workflows will contain an explanation that occurred (i.e. either P or M will have occurred, and either P or N will have occurred, and either P or M or N will have occurred, etc.)

Example 5 (Trace-Minimal Workflows). Of the six complete workflows discussed in Example 4, only PM and PN are trace-minimal.

One can see that workflows that are complete, task-minimal and trace-minimal are quite desirable, not only because they are very concise and thus can be verified easily, but also because one can be assured that the workflow will necessarily contain an explanation that is triggering the critical behaviour. In the examples above, if the workflow PM was chosen, a compliance officer observing that activity R took place could easily check to see whether either P or M was executed. If not, one could conclude that there was not a sufficient motivation for R to occur, and could choose to investigate the matter further.

5 Results

As a first step in the investigation of abductive workflow mining, we perform a few simple tests to get an idea of how the general idea can work by reducing the size of the workflow representation. To accomplish this, we employ a naive method for finding abductive workflows where we simply search the set of cases for strings of tasks that imply W'. This will help to accomplish the main goal of the paper, which is to investigate the potential of abductive mining in terms of its effectiveness in simplifying complex workflows.

We start by building a workflow model W for the entire set C. W is then used to discover traces in the model that correspond to a path through W'. An activity a_i that directly precedes this path and an activity a_o that directly follows it are then chosen to form a larger path that contains the path through W'. We then search the rest of the model to determine if there is a sequence that contains a_i and a_o, with activity not consistent with W' in between. If one is found, then a new a_i or a_o is selected, or perhaps a longer sequence of preceding or following activities is selected. If this is not the case, however, then executing a_i and a_o necessarily causes activity in W' to be performed in between, and thus this is an abductive trace. This process continues until a complete abductive model is found.

To get an idea of by how much the size of the workflow graph is reduced by simply considering the abductive model, we ran tests on five example log files that are packaged with the ProM process management software [5]. We used our own miner to build the workflow, and then we chose a single task to be the target and found the corresponding abductive model that explains the target. The average number of transitions and arcs produced for each of the original and abductive model are presented in Table 1. The data shows that, in these tests the abductive model was less than one quarter the size of the original in terms of the number of both transitions and arcs.

Table 1. Average size of the original and abductive workflow models for our workflow miner

	Original Workflow	Abductive Workflow
Number of transitions	156.2	37.2
Number of arcs	318.6	77.0

To show that the abductive model can score a significant reduction on workflows for other miners, we ran the tests on the α miner as well, a particularly tough one to reduce, since very few transitions are used in α-algorithm-mined workflows. The results for these tests are depicted in Table 2. The data shows that significant reduction took place, as the workflows were cut to less than half the size.

Table 2. Average size of the original and abductive workflow models for the α miner

	Original Workflow	Abductive Workflow
Number of transitions	12.0	5.6
Number of arcs	33.2	15.6

6   Conclusions and Future Work

In this paper we discuss a new approach to modeling workflow, referred to as abductive workflow mining. With this approach, small concise workflows can be modeled that are found to necessarily cause some activity in question to occur. This is especially useful in the area of compliance checking where there is a small number of critical tasks that need to be checked. The resulting smaller workflow means that errors are less likely, and can also be more easily understood by a human analyst that might be trying to make sense of what went wrong. We employ a rule mining approach that searches the log in an effort to construct task successor rules, which indicate which tasks follow which other tasks in the underlying process. These rules, which are initially represented in Kowalski form, are then converted to CNF, and binary resolution is used to determine the abductive explanations for the given critical activity. Tests on a very naive method for discovering abductive workflows show that the size of workflows can be significantly reduced.

While the technique performed well on a simple example presented in the paper, it is not guaranteed to find all abductive explanations. First, consider for example the transformed sets $C^+_{-W'} = \{PQw', QPw'\}$ and $C^-_{W'_o} = \{QPXw'_o\}$. Writing the clauses by preserving the order of tasks, we can see that P being executed before Q should necessarily imply W'. However, when considering these as clauses in binary resolution, PQ is treated no different from QP. And since QP does not necessarily imply W', no abductive explanations will be found. A second example of the technique's shortcomings arises in the transformed sets $C^+_{-W'} = \{PQRw', SQTw'\}$ and $C^-_{W'_o} = \{VTWw'_o\}$. Clearly, Q is an explanation for W'. However, the way the rules are written, Q is only specified to imply that R or T will be observed, and the observation of T does not imply W'. So Q will not be deemed an explanation by the resolution engine. Clearly there is much to be done to ensure that more (or all) explanations can be found. Accomplishing this task is the major focus of future work.

References

1. van der Aalst, W.M.P., van Dongen, B.F., Herbst, J., Maruster, L., Schimm, G., Weijters, A.J.M.M.: Workflow mining: A survey of issues and approaches. Data and Knowledge Engineering 47(2), 237–267 (2003)
2. van der Aalst, W.M.P., Weijters, A.J.M.M., Maruster, L.: Workflow mining: Discovering process models from event logs. IEEE Transactions on Knowledge and Data Engineering 16(9), 1128–1142 (2004)
3. Peterson, J.L.: Petri nets. ACM Comput. Surv. 9(3), 223–252 (1977)
4. Poole, D., Mackworth, A., Goebel, R.: Computational Intelligence: A Logical Approach. Oxford University Press, Inc., New York (1998)
5. ProM. The ProM framework (2007), http://is.tm.tue.nl/~cgunther/dev/prom/
6. Russell, S., Norvig, P.: Artificial Intelligence: A Modern Approach. Prentice Hall, Englewood Cliffs (1995)
7. van der Aalst, W.M.P., Rubin, B.F., van Dongen1, E.K., Günther1, C.W.: Process mining: A two-step approach using transition systems and regions. Technical report, Eindhoven University of Technology (2006)
8. van der Aalst, W.M.P., Weijters, A.J.M.M., Maruster, L.: Workflow mining: Discovering process mining models from event logs. IEEE Transactions on Knowledge and Data Engineering 16(9), 1128–1142 (2004)
9. van Dongen, B.F., Busi, N., Pinna, G.M., van der Aalst, W.M.P.: An iterative algorithm for applying the theory of regions in process mining. Technical report, Department of Technology Management, Eindhoven University of Technology (2006)
10. Weijters, A.J.M.M., van der Aalst, W.M.P.: Rediscovering workflow models from event-based data using little thumb. Integrated Computer-Aided Engineering 10(2), 151–162 (2003)
11. Wen, L., Wang, J., Sun, J.G.: Detecting implicit dependencies between tasks from event logs. In: Zhou, X., Li, J., Shen, H.T., Kitsuregawa, M., Zhang, Y. (eds.) APWeb 2006. LNCS, vol. 3841, pp. 591–603. Springer, Heidelberg (2006)

A Rule-Based Framework Using Role Patterns for Business Process Compliance

Akhil Kumar[1] and Rong Liu[2]

[1] Smeal College of Business, Penn State University, University Park,
PA 16802, USA
[2] IBM Research, 19 Skyline Drive, Hawthorne, NY 10532, USA
akhil@psu.edu, rliu@us.ibm.com

Abstract. In view of recent business scandals that prompted the Sarbanes-Oxley legislation, there is a greater need for businesses to develop systematic approaches to designing business processes that comply with organizational policies. Moreover, it should be possible to express the policy and relate it to a given process in a descriptive or declarative manner. In this paper we propose *role patterns*, and show how they can be associated with generic task categories and processes in order to meet standard requirements of internal control principles in businesses. We also show how the patterns can be implemented using built-in constraints in a logic-based language like Prolog. While the role patterns are general, this approach is flexible and extensible because user-defined constraints can also be asserted in order to introduce additional requirements as dictated by business policy. The paper also discusses control requirements of business processes, and explores the interactions between role based access control (RBAC) mechanisms and workflows.

Keywords: generic role patterns, constraints, rules, compliant business process, separation of duty, internal control, control policies, task categories, declarative approach, Sarbanes-oxley.

1 Introduction

Sarbanes-Oxley legislation in the United States and similar laws in other countries have highlighted the importance of making business processes secure. It has made it mandatory for top officers of organizations to certify that suitable controls are in place to guarantee that processes are secure. Section 302 of Sarbanes Oxley Act [16,17] requires that CEOs and CFOs must personally sign off on their company's financial statements, while Section 404 requires that appropriate processes and control must exist for all aspects of a company's operations that affect financial reports.

In this paper we discuss a framework for developing role patterns for business process compliance. For example, the requirements of an accounting application might state that: (1) an invoice must be approved before it is paid; (2) the goods must be received before the invoice is approved for payment; and, (3) the goods must be inspected before the payment is made. Similar needs arise in applications in patient care, immigrant processing, insurance claims, etc. When such a process is designed, an easy way is needed to do so.

N. Bassiliades, G. Governatori, and A. Paschke (Eds.): RuleML 2008, LNCS 5321, pp. 58–72, 2008.

The work in this paper is part of a larger framework consisting of four parts: *process patterns, role patterns, task categories* and *user-defined constraints*. A process pattern or workflow [1] is a generic pattern that specifies the ordering of tasks and subprocesses required for performing well-known functions in a business. Process patterns are quite general and apply in a variety of domains. Thus, a generic 'Order' process may be applied in various applications ranging from ordering a laptop computer to submitting a loan application. There are certain basic combinations of task categories that can be combined to create an *Order* process. Every order has to be *prepared*; it has to be *approved*; it has to be *submitted*; it has to be *received*; it has to be *paid for*; etc.

The second aspect of the framework is *role patterns*. *Roles* are standard designations or titles on the organization chart of any company. Role patterns are a way to restrict the roles that can participate in a process instance in terms of both the sequence in which the role can participate, and the number of times the role can participate in a process instance. For instance, an employee who fills in a travel request form for a business trip should get it approved from somebody in a different role, like the employee's manager. Role patterns can be associated with process patterns in order to enforce such compliance.

In this paper the focus is on designing processes using process patterns and then applying role patterns to them, and also to subprocesses and tasks within them. For space considerations, task categories and user-defined constraints are discussed only briefly in this paper. Section 2 provides basic background on the modeling and design of business processes along with an example, and also summarizes basic principles of internal control in a business. Next, Section 3 discusses a UML model for modeling processes, roles and users, and develops role patterns in the context of this model. It also shows how the role patterns can be implemented easily in Prolog. Then, Section 4 provides a discussion along with extensions, a high-level architecture for integration into an existing system, and related research. Finally, Section 5 concludes the paper.

2 Preliminaries

A business can be viewed as a collection of processes, and the robustness of these processes to a large extent is a crucial determinant of the success of the business. Business processes can be described using some simple constructs, and most workflow products provide support for these constructs. Four *basic constructs* that are used in designing processes are *immediate sequence, parallel, decision structure* (or *choice*) and *loop*, as shown in Fig. 1.

In general, business processes can be composed by combining these four basic patterns as building blocks. They can be applied to atomic tasks, e.g. Iseq(A,B) to indicate that tasks A and B are combined in sequence, or to subprocesses, e.g. Iseq(SP1, SP2) to indicate that subprocess SP1 and subprocess SP2 are combined in sequence. In Fig. 1(a), we use ISeq to indicate two tasks or subprocesses are in immediate sequence. Parallelism is introduced using AND-Split control nodes and parallel branches can be synchronized by AND-Join nodes at the end as shown in Fig. 1(b). A choice structure is shown in Fig. 1(c). In this pattern, the first OR construct, called OR-split, represents a choice or a decision point, where there is one incoming branch and it can activate any one of the two outgoing branches. The second OR construct is called an OR-Join because two incoming branches join here and there is one outgoing branch. Finally, it is also possible to describe loops in a process diagram by combining a pair of OR-split and OR-join nodes, such that one outbranch from the OR-Split node connects to an in branch of an OR-join node, as shown in Fig. 1(d). In this way, the patterns can be applied recursively to create a complete process.

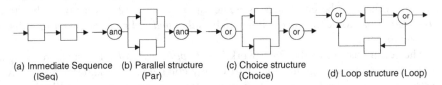

(a) Immediate Sequence   (b) Parallel structure   (c) Choice structure
       (ISeq)   (Par)   (Choice)   (d) Loop structure (Loop)

Fig. 1. Basic patterns to design processes

As a running example for this paper, Fig. 2 shows an *Administer Account Transfer* process. It starts with a **customer representative** receiving an account transfer instruction from a client. A **financial clerk** then checks if the details of this transfer instruction are complete, and, if so, gives an affirmative reply. If the instructions are incomplete, communication details regarding the invalid payment instruction are extracted. If the payment instruction is accepted, the transaction limit is checked by a **financial accountant**, and if it is within the limit then the availability of funds is tested by a **banking specialist**. If the transaction limit is exceeded, a request for authorization is made, and it should be approved by the finance manager, and then the transaction proceeds normally, i.e. appropriate accounting entries are created and applied to the required accounts, communication details are extracted from the accounting entries and the customer is notified. Finally, the system generates the communication details, the **senior finance manager** approves the transaction and the customer representative notifies the client.

Some observations about this process are as follows. First, this process is composed of individual tasks and subprocesses. Two key subprocesses are shown inside dotted-line boxes in Fig. 2. These are accounting entry and authorization, and each is composed of atomic tasks. In general, process design can be simplified by breaking down a process into subprocesses so that each subrpocess can be designed independently. Moreover, an OR-split node represents a choice or a decision point. Thus, both t3 (validate transfer instructions) and t4 (check transaction limit) are decision points. A parallel structure is introduced in the subprocess "accounting entry" where, when the funds availability test passes, two entries, one for business accounting (t10a) and the other for any fee related to this transaction (t10b), are created in parallel and then merged in an AND-Join node. In the example of Fig. 2, we also show the role that performs a task at the side.

2.1 Basic Control Requirements and Principles

At the outset it should be noted that control requirements are necessary in almost any business process where exchange of money or goods is involved. Moreover, this would apply regardless of whether a system is fully automated, partially automated, or entirely manual. In the automated case, the computer system should have been tested thoroughly before hand to make sure it operates correctly. In the manual case, the human worker must have the appropriate qualifications and authority to perform the task. In all cases, appropriate controls must be in place to prevent fraud or abuse of authority. Therefore, in this section we introduce basic control principles in a general manner. Subsequently, we will discuss how these principles can be operationalized.

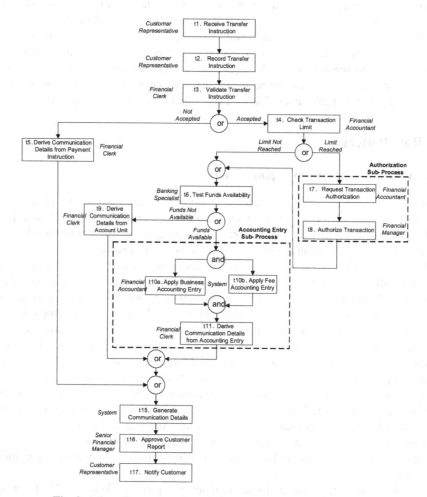

Fig. 2. A formal representation of an account transfer request process

The first standard principle of control is that a *requester* and *approver* for any task must be different [16]. This is the simplest situation of a separation of duties. Thus, a manager cannot approve his or her own expense claim for a business trip, but can do so for everybody else in the department. A further extension is the **"three-eyes"** rule. This requires separation of *custody, approval* and *recording* functions. Thus, for receipt of goods from a vendor, physical custody is kept by one person, the approval of the receipt is given by another, and the recording of the receipt is done by a third person. This ensures that the person receiving the shipment does not record it incorrectly. By separating receipt from recording, chances of fraud at delivery time are reduced. Moreover, there may be an additional requirement that the three individuals performing these tasks must be from different roles (say, inventory clerk, department supervisor and accounting clerk, respectively). A further extension of this is the **"four-eyes"** rule which may require that for an order, the *requester, authorizer, preparer* of payment and the one that *releases* it, all be different individuals. For extra-sensitive transactions, multiple approvals may be required instead of one at each stage, for example by having two approvers (say, a manager and a VP) instead of one.

In addition to separation of role requirements, ordering restrictions on roles may also be imposed. Thus, it may be necessary that a superior role (such as a manager or vice-president) may perform a task after a subordinate role (such as engineer).

3 Role Patterns

3.1 A Basic Model for Developing Patterns

A process description includes the tasks that are performed in a process and the order in which they are performed. Along with a process description, it is important to provide information about who will perform a task. In general, this is accomplished by means of *roles*. A role is an organizational position that is qualified to perform certain tasks. Thus, in order to perform a task, a user must belong to a certain role that is qualified and authorized for it. Examples of roles in any typical organization are manager, director, VP, secretary, CEO, etc. The permissions or "power" of a person depends on her role. Thus, a department manager may approve travel requests for the employees in his or her department, while the human relations manager may approve leave requests, and the technology manager may approve requests for computer purchases. For some requests multiple approvals may be required. Moreover, some individuals may also hold multiple roles, e.g., a person may be a department manager and also an electrical engineer. However, in many situations an individual can play only one role for a particular process instance. Thus, a manager in the role of *department employee* while submitting her travel expenses cannot later assume the role of *department manager*, and approve her own expenses. This is akin to the notion of separation of duties discussed earlier, and a scenario like this is forbidden by the organization policy in most companies.

Fig. 3 shows a basic model for developing role patterns. This is a UML model showing that tasks are subtypes of processes, and processes combine together using relationships like sequence (S), parallel (P), choice (C) and loop (L) to create larger processes. A process needs roles and roles have users in them. A process has attributes like:

```
min_num_roles: minimum number of roles that must participate in a task
min_role_level: at least one role must be at min_role_level
```

Similarly, permissions are attached to process-role and process-user binary associations by means of suitable attributes as follows:

```
role(user)_exclusions: other role (user) that must be excluded with this
role(user)_inclusions: other role (user) that must be included with this
max_tasks_role(user): maximum number of tasks a role (user) can perform
```

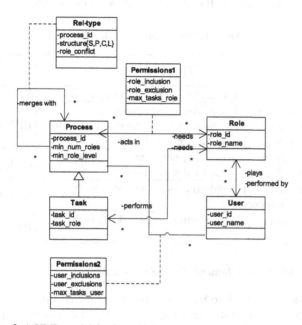

Fig. 3. A UML model for formalizing treatment of compliance issues

Finally there is a Rel-type class associated with the merge relationship between a pair of processes. This class has a Boolean attribute called *role_conflict* that determines whether a role can participate in both the merging processes. This class also has an attribute to determine the merge pattern (e.g. sequence, parallel, choice, or loop) between two merging processes.

3.2 Role Patterns

Our role patterns were inspired by this basic model and are shown in Table 1. They are a means of enforcing organizational policy on processes and can apply to tasks, subprocesses, and also complete processes. Pattern RP1 enforces participation of a minimum number of roles in the process. This corresponds to the attribute *min_num_roles* in the Process class in Fig. 3. Pattern RP2 restricts two tasks from being done by the same role if they have a certain type of relationship between them. Hence, if two tasks are in parallel, they may be required to be done by different roles. RP3 prevents two tasks from being done by the same role if they belong to different subprocesses. Thus, if a role takes part in an ordering subprocess, it may not also take part in a payment subprocess. RP4 requires that at least one task in a process or sub-process be done by a given role *min_role* (or higher). Thus, in a loan approval process there may be a requirement that a vice-president must approve a loan. This corresponds to the attribute *min_ role_level* of the Process class in Fig. 3. Finally, RP5 restricts the number of tasks a role can perform in a process.

These patterns are discussed further in the context of the running example below.

Table 1. Proposed role patterns (RP)

RP #	Role Pattern (RP) Description	Formal Expression
1	A (sub) process p must contain at least N unique roles.	RP1(p, N)
2	No pair of tasks with the lowest level common relationship *Rel* can be done by the same role in a (sub) process p, *Rel* $\in$ {Iseq, Par, Choice, Loop}	RP2(p, Rel)
3	No task pair from a pair of different sub-processes, say $sp1$ and $sp2$, can be done by the same role.	RP3($sp1,sp2$)
4	At least one task in (sub) process p must be done by *min_role* or higher.	RP4(p, min_role)
5	A role r can perform a maximum of N tasks in (sub) process p.	RP5(p, r, N)

3.3 An Approach for Implementation of Role Patterns

Next we discuss an approach to implement these role patterns in a rule language. We have chosen Prolog [9] to illustrate the approach. We first define base predicates and then use them to create the role patterns. The base predicates are as follows:

```
task(t): t is a task
role_assign(t,r,proc): assign role r to task t for process proc
child(p1, sp1): child of process p1 is subprocess sp1
parent(p1,sp1): p1 is a parent process of sp1
anc(sp1, t1,t2): first common ancestor of tasks t1, t2 is subprocess sp1
contain(p1,sp1); contain(p1,t): p1 contains sp1; p1 contains task t
          (note: a process also contains itself)
merge(sp1,sp2,p12,rel): subprocesses sp1, sp2 are merged into p12 in
                        'rel' relationship
role_occurs(t,r): role r occurs in task t
doer(t,u,proc): User u is the user or worker who performs task t for
                process proc.
```

We also use three built-in predicates as follows:

```
setof(X,P,S): built-in predicate is true if and only if S is the set of
              all instantiations of X that make P provable.
length(S,M): built-in predicate is true if the number of elements in set
             S equals M.
member(X,S): built-in predicate is true if element X is in set S
```

The realization of the role patterns is shown in Table 2. They are expressed in such a way that if a role pattern is true, then it means there is a violation. Next we explain the pattern rules and illustrate them in the context of the example of Fig. 2.

RP1(*proc*, 5). This pattern requires that in process *proc* any role cannot do more than 5 tasks. The rule creates a set (called *Rset*) of all tasks in a process proc in which every role R participates. The length of this set is determined, and if it is more than N, then the left-hand side of the rule becomes true, indicating a violation.

RP2(*"accounting entry"*, *"par"*). A pair of tasks with a parallel *(par)* relationship between them cannot be done by the same role. Thus, in the accounting entry subprocess if two tasks are in parallel then the roles must be different. This considers all subprocesses of a process, and examines all pairs of tasks that have the subprocess as their lowest level ancestor. Then if the merge relationship at the subprocess is of type 'Rel' (in this example parallel or par), then a violation occurs.

Table 2. Implementation of basic role patterns

Rp1(Proc ,N) :-	setof(R, role_occurs(Proc,R),Rset), length(Rset, M), M > N.
Rp2(Proc, Rel) :-	contain(Proc, SP1), anc(SP1, T1,T2), T1 ≠ T2, merge(_, _, SP1, Rel), role_assign(T1, R, Proc), role_assign(T2, R, Proc).
Rp3(Proc1, Proc2):-	contain(Proc1, T1), contain(Proc2, T2), T1 ≠ T2, role_assign(T1, R, Proc), role_assign(T2, R, Proc).
Rp4(Proc,Min_role):-	setof(R, role_occurs(Proc, R),Rset), not(member(Min_role, Rset)).
Rp5(Proc,R,N) :-	contain(Proc, T), setof(T,role_assign(T, R, Proc), Tset), length(Tset,M), M >= N.

RP3("authorization", "accounting entry"). No task pair from a pair of different subprocesses, say *sp1* and *sp2*, can be done by the same role. Thus, no task pair from authorization and accounting entry subprocesses can be done by the same role. The rule examines all pairs of tasks in the two subprocesses, and if any pair is done by the same role, then a violation is produced.

RP4(*proc*, "senior financial manager"). This pattern ensures that at least one task in a process or subprocess is done by a certain given role. In this example, we are asserting that the senior financial manager role must perform at least one task in the process *proc*. The rule creates a set of all roles that participate in the process *proc*, and then the built-in predicate member checks if the desired role is present in this set.

RP5(*proc*, "senior financial manager", 3). This is a way to limit the participation of a role in a process. The senior financial manager role can do no more than 3 tasks in the process *proc*. The rule considers all tasks in a process proc and then makes a set of all tasks performed by the given role. Then if the number of tasks is more than the limit, it means the role pattern is violated.

Next we consider some additional patterns.

3.4 Additional Role Patterns

In a similar way it is also possible to add rules for ensuring inclusion and exclusion of roles. For role exclusion we assume there is a predicate *role_exclude* that captures all pair of incompatible roles. Then a *Role_exclude1* predicate can be written as:

```
Role_exclude1(Proc,R1,R_excl)   :-
                    role_occurs(Proc, R1),
                    role_exclude(R1, R_excl),
                    role_occurs(Proc, R_excl).
```

Similarly, a *role_include1* predicate can be created as:

```
Role_include1(Proc, R1,R_incl)   :-
                    role_occurs(Proc, R1),
                    role_include(R1, R_incl)
                    not(role_occurs(Proc, R_incl)).
```

Above, role_exclude1 and role_include1 are predicates that contain information about roles that should be, respectively, excluded and included along with a given role, say, R1. Another common requirement that may arise is that a user may in general be allowed to perform multiple roles, but not all in the *same instance* of a process. Thus, if a predicate *doer(u, r, proc)* is used to describe that user *u* can play role *r* in process *proc*, then, we can restrict the number of roles that this user can perform (say, to N) with a *restrict_user_role* pattern:

```
Restrict_user_role (Proc, R, N) :-
          contain(Proc, T),
          setof(T,assign_role(T, R) ∧ doer(T, U, Proc), Tset),
          length(Tset,M), M > N.
```

3.5 An Overall Approach

Above we have developed a methodology to systematically manage controls in business processes. The main features of our approach are:

1. Basic *process patterns* are used to describe processes
2. Basic *role patterns* are used to describe control requirements.
3. The role patterns are associated with a process at different levels of granulatiry (i.e. whole process, subprocess, etc.) as per the business policies.
4. The patterns are implemented in a logic-based software application (such as Prolog).
5. Before making any task assignment to a role, the software performs checks and disallows certain tasks if they violate the control requirements.

4 Discussion, Extensions, Architecture and Related Work

If businesses have to incorporate security in their business processes in order to achieve, say, Sarbanes Oxley compliance, then it will have to be done in a systematic manner, not as an afterthought. Hence, there is a need for formal frameworks for ensuring that appropriate controls are in place. In this paper we have described a rule-based approach for implementing controls and ensuring compliance in a business process. The role patterns represent a generic way to associate rules with a process at different levels of granularity. The declarative nature of the approach makes it flexible. It is easy to change and add/delete a role pattern or apply it at a different level of granularity. Prolog was chosen for its simplicity as way to illustrate role patterns. Other possible ways to implement role patterns are OCL [27] and LTL [18].

4.1 Extensions

Two other aspects of our framework are *task categories* and *user-defined constraints*. They are discussed only briefly here for space considerations.

In general, tasks in a business process can be classified into certain generic task categories. As a starting point for this work, we have developed 10 categories (see Table 3). These categories were inspired by the Financial Services Workflow Model of IBM Information FrameWork (IFW) [21], a comprehensive set of banking specific business models that represent best practice in banking industry. The role patterns discussed in Section 3 can be applied to these categories of tasks, rather than to individual tasks. Thus, it would be possible to state that a role cannot do more than two tasks that fall in the requisition category, or that the same user cannot do two successive tasks that are in the approve category.

Table 3. Generic task categories

Task category	Description
Prepare	Make something ready for use
Record	Note, enter into system, store in database
Approve	Accept, reject, decide, signoff
Requisition	Request, ask, initiate, order
Transmit	Notify, provide, deliver, send payment, goods etc. (outside the organization).
Acquire	Receive, obtain
Administer	Manipulate, move, inquire, search
Inspect	Test, evaluate, check
Suspend/Terminate	Hold, finish, complete, stop temporarily
Report	Prepare a report, or any kind of output

A second extension of the framework is the addition of user-defined constraints. Some standard constraints are captured in the task and role patterns; however, more specialized and fine-grained constraints can also be added by allowing users to define their own constraints. Examples of such constraints would be: additional signatures on large payments, added approvals for new vendors, end of day review of all large payments, strong physical and system access controls, etc. In some cases, if these constraints occur frequently, they could be treated as new role patterns, thus making this framework extensible.

4.2 An Architecture for Adopting Role Patterns

Fig. 4 gives a high-level architecture for integrating role patterns into an existing workflow system. We propose the addition of a new module called the security requirement modeling module, which would allow users to describe their compliance needs. These needs are converted into role patterns which are stored in the database. Moreover, the process description schema in the existing process modeling module is also translated into process facts including task definition, role assignment and process patterns, and stored in a database. This requires a translation program that can examine the current process schema and convert it into the new format of the process patterns. When a task assignment is made by the business process modeling module, it would call a query engine that would run a query against the database to ensure that none of the constraints is violated. If there is a violation of any role pattern, then the query engine will prevent the assignment from being made and explain the violation to the user. Moreover, during the process execution, the process execution engine can also query the database to ensure that any changes to the role assignment made dynamically during the runtime (e.g. delegation or reassignment) still comply with the role patterns. Finally, security control at the user level (e.g., any user cannot play roles as both requester and approver) can also be added to the database and can directly constrain the process execution.

Moreover, the translation from the process description in an existing workflow system to the process patterns does not have to be very precise. Workflow systems offer a variety of constructs or patterns to capture complex coordination requirements [1]. For purposes of enforcing the role patterns, and in the interest of compatibility, such precise translation of each construct is not required. Since the role patterns refer only to four basic process patterns, the translation at a minimum only needs to capture the tasks in processes (and their subprocesses) with respect to just these patterns. Therefore, the translation can be done efficiently in an approximate manner and the role patterns can still be verified.

4.3 Related Research

Business process modeling allows business analysts to formally define a process to reflect the inner workings of a business. This exercise is formalized by using a standard methodology and a tool for business process modeling. There are several tools such as IBM Websphere Business Modeler [20] that allow a visual model of a business process to be built. However, most tools do not provide adequate support for security and this is often added in a piecemeal and rather adhoc manner.

Some recent, closely related research to ours is discussed in [13,14]. Here the authors have developed a declarative approach for process modeling using the SBVR (Structured Business Vocabulary and Rules) vocabulary and created a new framework. The vocabulary is supported by a model and allows process modeling and specification of access constraints in an English-like language. They also support defeasible logic [4,26] which is a non-monotonic logic and can work with a set of inconsistent constraints. Another approach for handling compliance inspired by defeasible logic and deontic logic [5] is discussed in [28]. These logics are more advanced than Prolog, and are based on notions of permissions, obligations and prohibitions. They are applied in the context of the Business Contract Language (BCL) [15,23] where the focus is on how to proceed when one party fails to meet its obligations. In such situations, the negligent party is obliged to perform some other actions in order to make certain amends for its failure as specified in BCL. A shortcoming of Prolog is that it does not allow description of such scenarios easily. In [2], the authors have used temporal logic expressions to check whether a log corresponds to constraints. They express their constraints in LTL [18] and use a tool called LTL checker to verify if certain desired properties of the log are satisfied. There is a tradeoff between complexity and expressive power, and an in-depth comparison across approaches is a subject for future research.

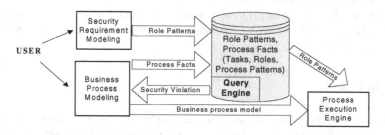

Fig. 4. An architecture for adding security in existing workflow engine

Prior research has looked at the issue of security from various perspectives. However, the stream of security related research that is relevant here relates to role based access control (RBAC) and was pioneered by Sandhu [29]. The basic RBAC framework consists of three entities: *roles, permissions* and *users*. Roles (such as manager, director, etc.) are assigned permissions or rights (to hire an employee, approve a purchase, etc.) and users (Joe, Lin, Sue) are associated with roles. Thus, users acquire certain permissions to perform organizational tasks by virtue of their membership in roles. The notion of separation of duties [22,30], although it preexisted in accounting and control systems, also reemerged in the context of RBAC as the idea that if task 1 is performed by role A, then task 2 must be performed by role B, and membership of these roles must not intersect. There are two types of separations of duty: *static* and *dynamic*. In recent years, RBAC has become the preferred access control model for most business enterprises. This framework allows association of roles with tasks, and only users that belong to a certain role can perform certain tasks. This is a useful framework that has now been widely adopted in popular database management systems from IBM and Oracle.

Some related work on specification and enforcement of role-based authorizations in workflow systems is discussed in [7]. The main focus of this work is on enforcement of constraints at run-time. The authors develop algorithms to check whether, given a combination of tasks and users, it is possible to find a task assignment that will satisfy the collection of constraints and available users. A formal model called W-RBAC for extending RBAC in the context of workflows using the notions of case and organizational unit is described in [31]. A system architecture for enforcing

RBAC in a Web-based workflow system is given in [3]. The approach in [8] is based on the notions of conflicting roles, conflicting permissions, conflicting users and conflicting tasks. More sophisticated algorithms for enforcing separation of duties in workflows are developed in [24]. Our work differs from and also complements these works in that our focus is on role patterns, and our goal is to give end users the ability to associate one or more patterns with processes. Moreover, we have a more sophisticated process model, resulting in a tighter integration between the process model and the compliance model. Furthermore, our process model is not hardcoded into the constraints, unlike in previous models, and thus it offers greater flexibility for associating tasks in a process with role patterns.

Finally, another stream of prior work that informs our research is the literature on basic financial control principles, particularly as it relates to the recent Sarbanes-Oxley legislation [6,10,16,17].

5 Conclusions

The focus of enterprise business process management lies in automating, monitoring and optimizing the information flow in an organization [11]. Moreover, internal controls are most effective when they are built into the enterprise infrastructure [10]. Therefore, internal controls must be tightly linked to business processes. Companies are also starting to realize the strategic value of making automated processes a part of daily business practice [6]. The focus of this paper is on creating a framework to embed controls into the processes of an organization. The main elements of this framework are process patterns and role patterns. We showed how a user can describe a process in a hierarchical manner using simple process patterns such as sequence, parallel, choice and loop, and then associate one or more of standard role patterns with it to create a compliant business process. Moreover, this can be done in a declarative manner and at different levels of granularity of the process. Although we did not discuss them at length, task categories and user-defined constraints can also be added for greater flexibility and for meeting special requirements.

The key advantages of this approach are that it is generic, easy to use and flexible. Hence, it offers simplicity and practical value. The role patterns that capture common control requirements can be associated with process patterns quite easily. Finally, the role patterns are not hardcoded and can also be extended. In future work, we expect to implement this framework and test it in a real environment. We would also like to add temporal extensions, perhaps in LTL [18]. For instance, consider a policy like, "A manager cannot approve any requests until she has been in the manager role for 6 months." Here an individual may be in the manager role, but she may still not perform certain tasks. To handle such situations, one possibility is to create a special role called 'New Manager' and not associate it with certain tasks. However, more flexible ways of dealing with such situations are required. Finally, it would also be useful to consider issues of delegation [31] (i.e. can a role delegate its tasks to other roles?), and explore how organization policy on delegation could be incorporated into the process securely. Lastly, a comparison between a monotonic approach such as ours and non-montonic ones such as in [13,14,28] will also be useful.

References

1. van der Aalst, W.M.P., et al.: Workflow patterns. Distributed and Parallel Databases 14(3), 5–51 (2003)
2. van der Aalst, W.M.P., Beer, H., van Dongen, B.: Process mining and verification of properties: An approach based on temporal logic. In: Meersman, R., Tari, Z. (eds.) OTM 2005. LNCS, vol. 3760, pp. 130–147. Springer, Heidelberg (2005)
3. Ahn, G.-J., et al.: Injecting RBAC to secure a web-based workflow system. In: Fifth ACM Workshop on Role-Based Access Control, Berlin, Germany (July 2000)
4. Antoniou, G., et al.: Representation results for defeasible logic. ACM Trans. Comput. Log. 2(2), 255–287 (2001)
5. Antoniou, G., Dimaresis, N., Governatori, G.: A System for Modal and Deontic Defeasible Reasoning. In: Australian Conference on Artificial Intelligence 2007, pp. 609–613 (2007)
6. Berg, D.: Turning Sarbanes-Oxley Projects into Strategic Business Processes. Sarbanes-Oxley Compliance Journal (November 2004)
7. Bertino, E., Ferrari, E., Atluri, V.: The specification and enforcement of authorization constraints in workflow management systems. ACM Trans. Inf. Syst. Secur. 2(1), 65–104 (1999)
8. Botha, R.A., Eloff, J.H.P.: Separation of duties for access control enforcement in workflow environments. IBM Systems Journal 40(3) (2001)
9. Clocksin, W.F., Mellish, C.S.: Programming in Prolog. Springer, New York (1987)
10. Committee of Sponsoring Organizations. Internal Control – Integrated Framework, http://www.coso.org/publications/ executive_summary_integrated_framework.htm
11. Ferguson, D., Stockton, M.: Enterprise Business Process Management - Architecture, Technology and Standards. In: Business Process Management, Vienna, Austria, pp. 1–15 (2006)
12. Gamma, Erich, et al.: Design Patterns: Elements of Reusable Object-Oriented Software, hardcover. Addison-Wesley, Reading (1994)
13. Goedertier, S., Mues, C., Vanthienen, J.: Specifying Process-Aware Access Control Rules in SBVR. In: Paschke, A., Biletskiy, Y. (eds.) RuleML 2007. LNCS, vol. 4824, pp. 39–52. Springer, Heidelberg (2007)
14. Goedertier, S., Vanthienen, J.: Declarative Process Modeling with Business Vocabulary and Business Rules. In: Proceedings of Object-Role Modeling (ORM 2007) (2007)
15. Governatori, G., Milosevic, Z.: A Formal Analysis of a Business Contract Language. Int. J. Cooperative Inf. Syst. 15(4), 659–685 (2006)
16. Green, S.: Manager's Guide to the Sarbanes-Oxley Act: Improving Internal Controls to Prevent Fraud. Wiley, Chichester (2004)
17. Haworth, D., Pietron, L.: Sarbanes-Oxley: Achieving Compliance by Starting with ISO 17799. Information Systems Management (Winter 2006)
18. Holzmann, G.: The Spin Model Checker. Addison-Wesley, Reading (2003)
19. Huang, W.-K., Atluri, V.: Secureflow: a secure web-enabled workflow management system. In: Proceedings of the Fourth ACM Workshop on Role-Based Access Control, pp. 83–94 (1999)
20. IBM Websphere Business Modeler (WBM), Version 6, http://www-306.ibm.com/software/integration/wbimodeler/
21. Information FrameWork (IFW), IBM Industry Models for Financial Services, http://www03.ibm.com/industries/financialservices/doc/ content/bin/fss_ifw_gim_2006.pdf

22. Kuhn, D.R.: Mutual Exclusion of Roles as a Means of Implementing Separation of Duty in Role-Based Access Control Systems. In: Proceedings 2nd ACM Workshop on Role-Based Access Control, Fairfax, VA, pp. 23–30 (October 1997)
23. Linington, P., et al.: A unified behavioural model and a contract language for extended enterprise. Data Knowl. Eng. 51(1), 5–29 (2004)
24. Liu, D., et al.: Role-based authorizations for workflow systems in support of task-based separation of duty. J. Syst. Softw. 73(3), 375–387 (2004)
25. Nagaratnam, N., et al.: Business-driven application security: From modeling to managing secure applications. IBM Systems Journal 44(4) (2005)
26. Nute, D.: Defeasible logic. In: Handbook of logic in artificial intelligence and logic programming: Nonmonotonic reasoning and uncertain reasoning, vol. 3. Oxford University Press, Inc., New York (1994)
27. Object Management Group (OMG), Object Constraint Language (OCL),
    `http://www.omg.org/technology/documents/`
    `modeling_spec_catalog.htm`
28. Sadiq, S., Governatori, G., Namiri, K.: Modeling Control Objectives for Business Process Compliance. In: BPM 2007, pp. 149–164 (2007)
29. Sandhu, R., Coyne, E., Feinstein, H., Youman, C.: Role-based access control models. IEEE Computer 29(2), 38–47 (1996)
30. Simon, R., Zurko, M.E.: Separation of Duty in Role-Based Environments. In: Proceedings of the 10th Computer Security Foundation Workshop, Rockport, MA, June 10–12, 1997, pp. 183–194 (1997)
31. Wainer, J., Kumar, A., Barthelmess, P.: DW-RBAC: A Formal Security Model of Delegation and Revocation in Workflow Systems. Information Systems 32(3), 365–384 (2007)

Detection of Suspicious Activity Using Different Rule Engines — Comparison of BaseVISor, Jena and Jess Rule Engines

Jakub Moskal[1] and Christopher J. Matheus[2]

[1] Dept. of Electrical and Computer Engineering, Northeastern University, Boston, MA USA
jmoskal@ece.neu.edu
[2] VIStology, Inc., 61 Nicholas Road, Framingham, MA USA
cmatheus@vistology.com

Abstract. In this paper we present our experience working on the problem of detecting suspicious activity using OWL ontologies and inference rules. For this purpose we implemented partial solutions using three different rule engines - BaseVISor, Jena and Jess. Each of them required different levels of effort and each had its strengths and weaknesses. We describe our impressions from working with each engine, focusing on the ease of writing and reading rules, support for RDF-based documents, support for different methods of reasoning and interoperability.

1 Introduction

Every merchant and naval vessel is required to periodically broadcast information about itself. Such information, handled by the Automatic Identification System (AIS) [1], consists of the vessel's location, speed, bearing, etc. Maritime activity of vessels is also tracked by other tracking-enabled vessels and static sensors that are located in range of the particular object. Therefore, information about one vessel may be coming from multiple sources and fusing them together is a challenge for reasons including incomplete or contradictory information. Current tracking systems fuse incoming information and through calculations estimate vessels' current location, speed and bearing. Suspicious activity, however, must for the most part be detected by human operators. The goal of our ongoing work is to design systems that can help perform this task automatically.

Our solution uses the OWL Web Ontology Language [2] to represent tracking information and rules to detect suspicious activity based on that information. One of the challenges was to design a system that could be used in a real-time environment. In our solution, we process data in chunks, each time running a new instance of the rule engine, so that only the recent snapshot is considered and there is no need to retract historical data. The system was implemented using three rule engines - BaseVISor, Jena and Jess; each required different amounts of effort and yielded different performance. The remainder of the paper describes our experience and impressions from working with each of them.

N. Bassiliades, G. Governatori, and A. Paschke (Eds.): RuleML 2008, LNCS 5321, pp. 73–80, 2008.
© Springer-Verlag Berlin Heidelberg 2008

1.1 Ontologies

As mentioned earlier, OWL ontologies are used to capture tracking information. An ontology is a representation of a set of terms within a domain and the relationships between those terms. OWL ontologies enable representing knowledge in a way that it can be processed by applications, not just by humans, because it explicitly represents the meaning of its terms. For that reason it is not only machine-readable, it is machine-interpretable, which suits our needs for automatic detection.

In our example, we designed two OWL ontologies[1]:

1. *Common Data Model(CDM)*, shown in Figure 1, includes basic vessel tracking information. `ObjectState` includes information about an object's location, bearing and speed at a particular time, this information is marked with some level of belief, here called `Confidence`. Each `Object` may have multiple aliases.
2. *SuspiciousActivity*, shown in Figure 2, represents the taxonomy of a few examples of suspicious activities that are to be detected. This ontology imports the *CDM* ontology and the two object properties, `susp:hadSuspiciousActivity` and `susp:suspectObjectState`, show the connection between them. Naturally, one could think of many other types of activities; here we only show a subset of a much larger suspicious activity ontology.

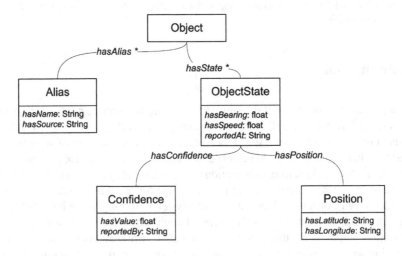

Fig. 1. The CDM ontology is trivial, however, it captures the necessary tracking information. The graph shows OWL classes and object properties.

1.2 High-Level Description of Rules

In order to demonstrate how suspicious activity detection could be aided by the use of rule engines, we designed a simple scenario. First, a sensor is suspected of being

---

[1] It should be noted that the ontologies shown in this paper are merely a subset of full ontologies which were not represented here for the sake of clarity.

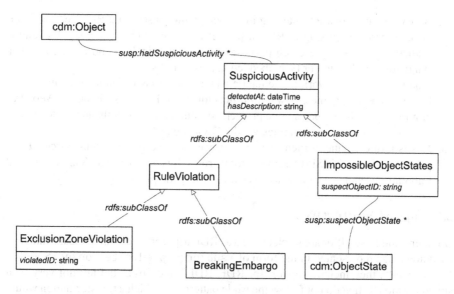

Fig. 2. The SuspiciousActivity ontology captures different kinds of activities that are considered suspicious. cdm:Object and cdm:ObjectState, OWL classes imported from the CDM ontology, are included in the graph to show how both ontologies are connected with each other.

compromised after its exclusion zone was violated, then some vessel moves in a direction different than it reports, being supported by the possibly jeopardized sensor. The vessel's tracking information is contradictory with other sources and it is considered suspicious. We used simulated data that followed this scenario and designed four rules to detect the suspicious activity. Each of them was implemented for all three rule engines that we tested. Below are their textual descriptions:

1. *Exclusion-Zone-Violation* - create an alert when a vessel violates the law by entering a restricted zone.
2. *Impossible-Object-States* - detect a situation when a vessel is reported in two different locations and the distance between them is impossible to cover within the time difference and under the assumption that all vessels have some arbitrary maximum speed.
3. *Report-Suspect-Sensor* - create an alert when an object was involved in some suspicious activity and is reported with contradictory information about its state.
4. *Report-Suspicious-Activity* - based on created alerts, report any suspicious activity.

The technical details of how these rules were implemented are omitted in this paper. We note that for each engine procedural attachments were implemented and this work describes our experience with it.

2   Rule Engines

The rules were implemented for the following three rule engines: BaseVISor [3], Jena [4] and Jess [5]. Below are their brief descriptions:

1. BaseVISor is a forward-chaining inference engine based on a Rete network optimized for the processing of RDF triples. It is developed by Vistology, Inc. and is available for free for research purposes. It is written in Java and can be run as a standalone application or be embedded in applications.
2. Jena is a Java framework for semantic web applications which also includes rule-based inference engines. It was first developed by HP Labs Semantic Web Programme, but it is an open source project. Although it has an embedded rule engine, it allows the use of other engines via the DIG[6] interface.
3. Jess is a rule engine written in Java that was inspired by the open-source CLIPS project, and uses a LISP-like syntax for its rules. It is developed by Sandia National Laboratories and is available at no cost for academic purposes.

2.1 Support for RDF/OWL

Both Jena and BaseVISor are implemented as RDF triple engines which means that they use knowledge represented in the form of <subject, predicate, object> triples. Consequently, they include support for RDF, RDFS and OWL, but are not very well suited for models that do not follow the triple pattern, for which a transformation would be necessary.

Jess on the other hand has a more general list-based knowledge representation with which the user can define arbitrary data structures using templates. As a result, unlike BaseVISor and Jena which can import RDF/S and OWL documents, Jess users have to define an RDF triple template and convert ontologies accordingly. The drawback of this approach is that it requires additional effort and there is no standard for doing so. In our work we used the OWL2Jess converter tool [7] to automatically transform OWL documents into Jess syntax.

2.2 Supported Methods of Reasoning

When working with inference engines, two main methods of reasoning are available - *forward* and *backward chaining*. Rules have a form of $LHS => RHS$. The left-hand side of the rule, LHS, is often called the *antecedent*, *body* or *if* part of the rule, and analogically, the right-hand side of the rule, RHS, is often called the *consequence*, *head* or the *then* part of the rule. Forward-chaining algorithms look for rules where the LHS is true and add RHS facts to the knowledge base, therefore it is data-driven. Backward-chaining tries to satisfy goals defined in the RHS, by checking if the LHS is true, and if it is not, the LHS is added to the list of goals. This method is therefore goal-driven.

There is a fundamental difference in the implementations of the three rule engines. BaseVISor and Jess are rule engines per se, implemented around the Rete network, an optimized version of forward-chaining [8]. While BaseVISor supports only forward-chaining, Jess also provides backward-chaining, which is effectively simulated in terms of forward-chaining rules. In order to use backward-chaining in Jess, users must explicitly state which facts(goals) and rules(triggers) can be used by that algorithm [5].

Jena itself is not a rule engine, but a framework for working with semantic web applications - it provides an API for handling RDF, RDFS, OWL and SPARQL. Reasoners are connected to it through interfaces and earlier versions of Jena were not released with reasoners at all. The latest version of Jena, however, comes with built in reasoners that support both methods of reasoning.

One could distinguish a third type of reasoning, *hybrid*, which is a combination of both forward and backward chaining. This type is supported by one of the built-in reasoners in Jena and in order to distinguish rules for each type, users use a different syntax, *LHS => RHS* and *LHS <= RHS* for forward and backward chaining, respectively. One could say that Jess also supports hybrid reasoning because rules can be marked for backward-chaining with (do-backward-chaining ruleName), although it is simulated and the engine works as a forward-chaining engine anyway.

2.3   Features Summary

The features described above are summarized in the Table 1. Based on the description of our problem, we required a forward-chaining inference engine due to the streaming nature of the data; all three engines meet this requirement. BaseVISor and Jena are best suited to our problem because of their support for RDF documents and data.

Table 1. Summary of the comparison of the used rule engines

Feature	BaseVISor	Jena	Jess
Support for RDF, RDFS, OWL	●	●	○
Forward-chaining implemented with Rete network	●	●	●
Backward-chaining	○	●	●
Hybrid reasoning	○	●	●
XML rule syntax	●	○	●

3   Reflections on Working with Each Engine

As shown in Figure 3, the overview of our system, the incoming tracking information was in XML format. It was later converted into OWL and fed into the rule engine. It is noted here that while one could solve our problem by dealing directly with the original XML data, we converted it to OWL/RDF in order to be able to leverage its inference capabilities (e.g., subsumption, property constraints, etc.). For Jess, an additional step was required to convert OWL into Jess format. Each engine had to be preloaded with the CDM and SuspiciousActivity ontologies, reasoning axioms, rules and procedural attachments (e.g., distance between two points). Ontologies were converted into Jess format from OWL, but this process was performed only once and the result was stored in a file.

Even though we were developing the same rules for different engines, in each case most of the time was devoted to writing the rules. Developing procedural attachments was necessary in each case, although different engines provided different means to do so. Below are our reflections from working with each of them.

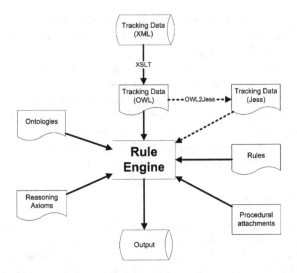

Fig. 3. Overview of the system. The dotted lines indicate additional processing necessary for running Jess engine which does not support importing OWL documents. Each engine must load rules, necessary procedural attachments, ontologies, axioms for inference other than that provided by the rules and the tracking data. The output is defined by the rules.

3.1   Writing Rules

A lot of time was devoted to debugging the rules, especially for errors that were not caught at compile-time or runtime. This comes from the fact that improperly written rules, though with proper syntax, do not provide any feedback to the user. For this reason, it is important that the rules are easy to read. One of the common mistakes was omitting the question mark in front of the variable name; both Jena and Jess treat such a clause as a plain literal and, in most cases, the particular rule would not fire. Having hundreds of lines of code makes the task of finding such a mistake challenging, especially since there might be multiple reasons for which a rule does not fire. In this case, the verbosity required by BaseVISor paid off, as an explicit `variable` attribute was never mistaken with a literal or a resource. Additional help was provided by the text editor which used syntax highlighting for XML.

Sometimes, in order to save time and make the rules easier to read, temporary variables were created, e.g. an intermediate result of some calculation. Each engine supports different styles of variable binding, which may significantly change the way rules are written. This applies especially to Jess, which does not allow binding in the LHS and in some cases requires writing additional rules to overcome this limitation. Such an approach may ruin the logical breakdown of the rules and make them harder to read. To keep the rules as close to each other as possible, instead of making temporary variables, we repeated the same calculations within the LHS of a particular rule.

Unlike BaseVISor and Jess, which provide an explicit `bind` function for arbitrary binding, Jena only allows implicit binding within procedural attachments. By convention, the last variable in a function may be bound, however, some functions do not bind variables at all. This approach has two major drawbacks. First of all, functions cannot be nested, because they do not return a value and one must use additional temporary variables. Secondly, one cannot tell from the syntax whether a variable is bound or not, thus forcing the user to consult the documentation. BaseVISor and Jess allow function nesting, although the deeper the nesting the harder it is to read.

3.2 Writing Procedural Attachments

Procedural attachments, also called functors, provide additional primitives that can be used in the rules. They are often used when performing tasks that would be hard to achieve with the declarative nature of the rules. Each engine provides a different set of built-in primitives, but they all provide functionality for printing, logical tests and basic math. Users are allowed to extend those sets by implementing their own procedural attachments and registering them with the appropriate instance of the engine.

In order to implement our rules, a few functions needed to be added. Since all engines were written in Java, the API that allows them to be extended is also available in Java. For this reason, part of the logic code could be reused. Out of the three engines, Jena provides the most control for handling errors, because it allows users to check if a particular parameter is a resource, literal, or a variable. Within BaseVISor and Jena, users parse function parameters to their expected types; any errors are handled by the engine.

The effort required for writing the procedural attachments differed across the rule engines. Jena distinguishes the behavior of a functor when it is used in the LHS and RHS; it can only return a boolean value when used in the LHS, so that it can be matched. On the other hand, functors in BaseVISor and Jess behave the same way in LHS and RHS; they may return any value that is part of the engine's datatype system, which later can be bound to a variable within a rule.

4   Summary

When choosing a rule engine, a lot of factors come into play and the best choice will most likely depend on the nature of the application. When the ease of writing the rules is important, Jena and BaseVISor are both a good choice. However, as much as writing the rules in BaseVISor is intuitive and does not require consulting the documentation too often, the rules can become large and it becomes more challenging to manage them when they take more space then one screen; BaseVISor's abbreviated syntax, however, does a nice job of helping to alleviate this issue. Jena rules are not only easy to write but are the easiest to read due to their succinct syntax. The appeal of reading and writing rules in Jena led us to create a converter that transforms Jena rules into BaseVISor syntax; one must only make sure that all used Jena procedural attachments are supported by BaseVISor. When changes to the rules were necessary, we would usually start with the Jena rules file and then transform them into Jess and BaseVISor.

When interoperability is important, XML support is usually highly desirable. When switching from Jena to other engines, users must write their own parsers for Jena's custom syntax. A possible solution to this problem would be using our converter tool to transform Jena rules to BaseVISor syntax, and then through XSLT, transform it to the required format. Jess, like in Jena, uses its own syntax, however, it provides conversion to the XML-based JessML. This makes it a better choice than Jena, but the problems with processing namespaces remain. BaseVISor not only uses XML-based syntax by default, but it also provides support for RuleML, which is designed for interoperability and has the added feature of supporting n-ary predicates.

Finally, when support for RDF-based documents is important, Jess is not the ideal choice, as it requires additional effort to convert them into Jess format. Although ontologies can be converted and stored in separate files, they take additional resources and require synchronization with the changes made in the ontologies. BaseVISor and Jena are both well suited for importing RDF-based documents, which probably makes them a better choice for Semantic Web applications in general.

While the nature of the problem addressed in this paper did not stress the systems enough to provide insight into their relative performance characteristics, previous tests [3] have shown BaseVISor to be significantly more efficient in time and space than Jess on RDF-based problems, which is not surprising given that BaseVISor uses a Rete algorithm optimized for RDF triples whereas Jess's Rete algorithm permits pattern matching on arbitrarily complex nested list structures.

References

1. Harre, I.: Ais adding new quality to vts systems. Journal of Navigation 53(03), 527–539 (2000)
2. W3C: Web ontology language (owl), http://www.w3.org/TR/owl-ref/
3. Matheus, C., Baclawski, K., Kokar, M.: Basevisor: A triples-based inference engine outfitted to process ruleml & r-entailment rules. In: Second International Conference on Rules and Rule Markup Languages for the Semantic Web, Athens, GA, USA, pp. 67–74 (November 2006)
4. Hewlett-Packard: Jena – a semantic web framework for java, http://jena.sourceforge.net/
5. Hill, E.F.: Jess in Action: Java Rule-Based Systems. Manning Publications Co., Greenwich (2003)
6. Bechhofer, S., Moller, R., Crowther, P.: The dig description logic interface. In: International Workshop on Description Logics (2003)
7. Mei, J., Bontas, E.P.: Reasoning paradigms for owl ontologies. Technical report, Department of Computer Science, Peking University and Institut für Informatik, Freie Universität Berlin (November 2004)
8. Forgy, C.L.: Rete: a fast algorithm for the many pattern/many object pattern match problem. Artificial Intelligence 19, 17–37 (1982)

A Rule-Based Notation to Specify Executable Electronic Contracts

Massimo Strano, Carlos Molina-Jimenez, and Santosh Shrivastava

Newcastle University, UK
{massimo.strano,carlos.molina,santosh.shrivastava}@ncl.ac.uk

Abstract. This paper presents a notation to specify executable electronic contracts to monitor compliance and/or enforcement of business-to-business interactions. A notable feature is that the notation takes into account the distributed nature of the underlying computations by paying due attention to timing and message validity constraints as well as the impact of exceptions/failures encountered during business interactions.

1 Introduction

Electronic contracts (e–contracts) can be used to regulate business interactions conducted online. By regulation we mean monitoring compliance and/or enforcement of business-to-business (B2B) interactions. E–contracts need to be made free from ambiguities that are frequently present in conventional contracts, where they are resolved by humans as the need arises. From our point of view a notation to specify an e–contract should possess four features: *usability and implementability, verifiability, expressivity, and exception handling*, detailed in [1]; yet we can briefly mention that usability and implementability suggest that the contract notation should offer a compromise between high level of abstraction and implementation details; verifiability suggests that the notation should be amenable to some form of verification to check for consistency; expressivity requires that the notation should be expressive enough to specify typical contractual clauses found in most practical applications.

As for exception handling, it should be possible to specify how to deal with the consequences of any exceptional situations that arise because of the inherently distributed nature of the underlying computations. First, there should be easy ways of specifying how to deal with any software and/or hardware related problems encountered during business interactions (e.g., unpredictable transmission delays, message loss, corrupted messages, semantically invalid messages, node failures, timeouts, etc.). Second, because B2B interactions typically take place between partners that are loosely coupled and in a peer-to-peer relationship, partners can sometimes get out of synchrony and perform erroneous, even mutually conflicting operations; an example is a buyer cancelling an order not knowing that the product has been shipped by the seller.

Our rule based language notation has been designed with all the above issues in mind and is particularly strong on exception handling. We have defined specific

N. Bassiliades, G. Governatori, and A. Paschke (Eds.): RuleML 2008, LNCS 5321, pp. 81–88, 2008.

exceptional outcomes (e.g., *technical failure, business failure*) on which numerous exceptional events that occur during business interactions can be mapped, and have suitably enhanced the familiar Event-Condition-Action (ECA) style rule notation. We also describe how such notation can be used for dealing with conflicting situations referred to above.

2   Contracts and Business Interactions

Our business scenario is shown in Fig. 1(a) where two enterprises (a buyer and seller) interconnected by a Message Oriented Middleware (MOM) trade with each other under an e-contract derived from a conventional one. The business interaction involves the shared execution of a set of primitive B2B *conversations* (*business activities* or *business dialogs*); examples are *purchase order conversation, notify of invoice conversation*, etc. At implementation level, we map each shared business operation onto a corresponding conversation, e.g., the operation *Issue Purchase Order* (PO), clause 1.1 is mapped onto *PO conversation* [2]. Thus, executing each Business Operation (BO) results in the execution of its corresponding conversation.

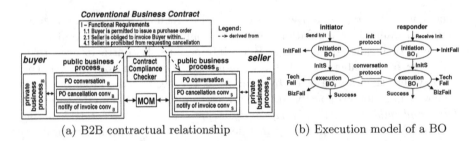

(a) B2B contractual relationship   (b) Execution model of a BO

Fig. 1. Definition and execution of a shared business process

A contract can be abstracted as a document that stipulates a list of clauses stating rights (R), obligations (O) and prohibitions (P) and associated constraints that two or more business partners are expected to honour. Informally, a right is something that a business partner is allowed to do; an obligation is something that a business partner is expected to do unless they are prepared to be sanctioned; finally, a prohibition is something that a business partner is not expected to do unless they are prepared to be sanctioned.

2.1   Execution Model for Business Operations

A primitive conversation representing a business operation is typically long lasting (taking hours to days to finish) and involves exchange of one or more electronic business documents (see industry standards such as RosettaNet partner interface processes and ebXML [3,4]). The industry practice is to use a MOM with transactional message queueing/dequeueing facilities for implementing conversations. Well known network protocol techniques are used to deal with problems such as lost and corrupted messages, but there are additional problems that need special attention [5]. Conversations have several timing and message validity constraints that need to be satisfied for their successful completion. A failure to deliver a valid message within its time constraint could cause mutually conflicting views of an interaction (one party regarding it as timely whilst the other party regarding it as untimely). A conflict can also arise if a sent message is delivered on time but not taken up for processing due to some message validity condition not being met at the receiver (so that the sender assumes that the message is being processed whereas the receiver has rejected it). Such conflicts will eventually lead to the parties having divergent views on the state of the shared business activity, causing them to take different, erroneous execution paths. It is important for e–contracts to incorporate mechanisms to deal with such cases. Our approach is discussed below.

We assume that for each BO_i initiator and responder are unique (see Fig. 1(b)); the initiation involves the execution of the *Init* protocol that produces a mutually agreed outcome which is either *InitFail* or *InitS*, respectively representing initiation failure and initiation success. BO_i is executed only when the outcome is *InitS*. Following ebXML specification [4], we assume that once a BO_i is started it always completes to produce at each side one of three possible events: *Success*, *BizFail* or *TecFail*, representing, success, business failure and technical failure, respectively. When a party considers that, as far as it is concerned, the conversation completed successfully, it generates the *Success* event. *BizFail* and *TecFail* events model the (hopefully rare) execution outcomes when a party is unable to reach the normal end of a conversation due to exceptional situations. *TecFail* models failures detected at the middleware level, such as a missing message. *BizFail* models semantic errors detected at the business level, e.g., the goods-delivery address extracted from the business document is invalid. Under normal conditions, both sides will produce identical outcomes; however, conflicting outcomes are possible, for example, one of them declares *Success*, whereas the other declares *BizFail*. It is worth emphasising that the conflicts mentioned above emerge at the execution level of conversations and are orthogonal to rule conflicts that could exist in a contract where some rules might contradict each other.

2.2   Contract Violations and Exception Handling

In practical business interaction, a partner, A, is forced to undertake exceptional actions, rather than follow the normal execution path, when A detects that the other partner, B, has breached (violated) a contract clause, (meaning B has failed to meet an obligation stated in the clause); exceptional actions usually involve imposition of some kind of sanction on B. Since a sanction is in fact an obligation that the offender is expected to honour, in the literature it is also termed *Contrary To Duty Obligation* (CTD) [6].

Exception handling can be made fairer if the underlying causes for the violations can be detected, and sanctions applied only when strictly necessary. In e–contracting, it is particularly important to distinguish violations caused by situations that arise primarily because of the inherently distributed nature of the underlying computations from those that are not and are mostly human related. Take a simple example: B fails to make a payment before the stipulated deadline. It makes sense to distinguish cases where the missing or delayed payment is owing to some infrastructure related problem (say the network was down) from cases where no such problems existed (so probably B was just late or deliberately avoiding payment); ideally, a sanction (such as a fine) should not be imposed on B under former cases, rather actions such as extending the deadline should be undertaken. As discussed in subsection 2.1, there is another source of exceptions caused by distributed computations that needs attention, namely, conflicting views on the state of a shared business activity that lead to erroneous executions, causing accidental breach. So for example, B executes a rightful *Cancel PO* operation that is viewed as successful at B but failed at A (the conversation generates *Success* at B, but, say, *TecFail* at A). Subsequent to that, B is no longer expecting to make a payment, but A is waiting for it; an accidental breach by B occurs. In such cases, it again makes sense not to impose sanctions on B, but incorporate some corrective actions. This philosophy underpins our exception handling approach.

Our execution model for business operations provides a uniform way of specifying exceptions caused by distributed computations. Wide variety of protocol related exceptions as well as those caused by timing and message validity constraints are mapped onto one of four outcome events (*InitFail* , *Success*, *BizFail* or *TecFail*) generated at each partner (see Fig. 1(b)). These events are sent to our execution environment (Section 2.3), where an event composer produces a single composite event as the outcome for a conversation, ensuring that a conversation is regarded as successful only if both parties have generated *Success*. More precisely: (a) identical outcome events are composed into a composite event of the same type; (b) if one of the outcome events is *TecFail* then the composite event is of type *TecFail*, irrespective of the type of the other event; (c) if one of the outcome events is *BizFail* and the other is not *TecFail*, then the composite event is of type *BizFail*. This classification enables concise specification of rules with violation cases that do not necessarily require imposition of sanctions. There would be no systematic way of dealing with such cases if exception management as suggested here is not incorporated. In our example in Section 4, clauses C7 and C8 illustrate this aspect further.

2.3   The Contract Compliance Checker

Our contract notation is based on the ECA paradigm and designed to be executed in our rule execution enviroment describing in [7] and consists of an event composer and Contract Compliance Checker (CCC). The event composer supports our execution model of Fig. 1(b) i.e., it receives the primitive events produced by the two parties, computes a single composite event and sends it to the CCC. The CCC is a neutral entity (conceptually located between the two interacting parties) that observes the ongoing business transactions and verifies their compliance with the contract. It contains a *time keeper* that generates timeout events when contractual deadlines expire; a *contract repository* that contains the rules; the ROP sets R_B, O_B, P_B and R_S, O_S, P_S that contain, respectively, the current rights, obligations and prohibitions of the buyer and seller; and the Relevance Engine (RE) executing the following algorithm: *1. Receive an event e from the event composer; 2. Analyse the contract repository and identify relevant rules for e; 3. For each relevant rule r, execute the actions listed in its right hand side.* Typically, these actions are $+=$ and $-=$ operations executed on the ROP sets.

3   Definition of the Notation

Role players: The keyword **roleplayer** defines a list of role players, e.g., **roleplayer** *buyer, seller* defines the two role players of our scenario.

Business operations: **businessoperation** defines a list of business operation types (e.g. **businessoperation** *POSubmission, InvoicePayment*).

Composite obligations: A composite obligation (defined with the keyword **compoblig**) is a tuple of obligations to be executed OR–exclusively by a deadline to satisfy the composite obligation, e.g., **compoblig** *RespondToPO(POAcceptance, PORejection)* defines the composite obligation to either accept or reject the PO stipulated in C2 of our example. The deadline is assigned at runtime, when the obligation is imposed.

Rules: A rule is defined with **when** *triggerBlock* **then** *actionBlock* **end**.

Trigger blocks: A *triggerBlock* is defined with *event* **is** *eventType [&& conditions]* and describes the match of an event to an event type, and a list of, possibly empty, conditions.

Conditions: A *condition* is a Boolean expression that evaluates:

- historical queries of the form **happened***(businessOperation, roleplayer [, outcome][, timeConstraint])*, where "*" can be used as a wildcard for all fields, while *outcome* defaults to *Success* and *timeConstraint* to the whole contract life if not specified.
- the identity of the originator or the responder of an event.
- the presence or absence of a business operation or composite obligation in the ROP sets of the participants;
- the outcome (*InitFail, Success, TecFail, BizFail*) of the executed business operation signaled by the event.

Action blocks: An *actionBlock* contains a list of $+=$, $-=$ and *pass* actions manipulating the ROP sets; *pass* has no effect, and $+=$ and $-=$ respectively add and remove rights, prohibitions and composite and non-composite obligations:

```
roleplayer.rights  += BizOp[(expiry)];        roleplayer.rights  -= BizOp;
roleplayer.prohibs += BizOp[(expiry)];        roleplayer.prohibs -= BizOp;
roleplayer.obligs  += BizOp(expiry);          roleplayer.obligs  -= BizOp;
roleplayer.obligs  += CompObl(expiry);        roleplayer.obligs  -= CompObl;
```

The actions of the left column, respectively, add the right, the prohibition and the obligation to execute the business operation *BizOp* and impose the composite obligation *CompObl*; the ones in the right column remove the right, the prohibition and the obligations to execute *BizOp* and *CompObl*.

Deadlines: In the examples above, *expiry* is a deadline constraint imposed on a role player to honour his rights, obligations and prohibitions; its absence indicates deadlines that never expire, lasting until the contract terminates. Notice that obligations with no deadlines are not of practical interest as their breach cannot be verified. We take the CCC clock's reading and the completion —as opposite to initiation— of an operation as the points of reference to declare a failure or success to meet a deadline.

Conditional statements: They can be used in the *actionBlock* of a rule, using *if* conditions **then** actions; [else actions;]

In an *actionBlock*, the keywords *Success, InitFail, TecFail, BizFail, Otherwise* represent the possible outcomes of the conversation; where *Otherwise* is a catch-all case; they are used to guard the actions of the rule, and can appear in any order and be grouped when they guard the same actions.

4   A Sample Contract

We will illustrate the use of our notations with the help of some clauses extracted from an hypotetical contract between a buyer (B) and seller (S). **C** and **R** stand for "clause" and "rule" respectively, so **C1** and **R2** respectively mean clause 1 and rule 2.

- **C1:** B has the right to submit a PO Mon–Fri, between 9 and 5 pm.
- **C2:** S is obliged to either accept or refuse the PO within 24 hrs. Failure to satisfy this obligation will abort the business transaction for an offline resolution.
- ...
- **C7:** If payment fails for technical or business reasons, B's deadline to respond to invoice is extended by 7 days and S gains right to cancel the PO.
- **C8:** B and S are obliged to stop the the business transaction upon the detection of 3 failures to execute the payment. Possible disputes shall be resolved offline.

Mapping of Clauses to Rules: In the simplest case the mapping between clauses and rules is one to one, yet the general case is M to N; all depends on the writing style. We show the rule representation of $C7$ and $C8$, only, as they illustrate the central ideas of our work; the full text of our contract example and its rule representation is provided in [1]. We start with the declaration of role players, business operations and obligations:

```
roleplayer buyer, seller;
businessoperation Invoice, Payment, POCancellation, TotalRefund,
compoblig RespondToInv(Payment, POCancellation);
```

R7 handles the seller's right to cancel a PO. If the buyer does not pay, its obligation to react to the invoice is removed, as steps will be taken to close the business transaction, which are not discussed here.

R7 corresponds to C7

```
when e is POCancellation
then # Seller cancels Purchase Order
  Success:
    if e.originator== seller
    && POCancellation in seller.rights
    && !happened(Payment, buyer)
    then
      buyer.obligs-= RespondToInv;
      # Conclude business transaction.
  Otherwise: pass;
end
```

R8 corresponds to C8

```
when e is Payment then
  Success: pass;
  Otherwise:
    if count(happened(Payment, buyer,
      InitFail))
    + count(happened(Payment, buyer,
      TecFail))
    + count(happened(Payment, buyer,
      BizFail)) >= 3
    then
      # Conclude business transaction.
end
```

R8 represents *C8* and restricts to three the number of unsuccessful attempts to make a payment. Fig. 2 shows four possible timelines of the *Payment* conversation (see C7 and C8). In the first scenario the payment succeeds in the first attempt within the seven day deadline (7d). In the second, it fails once due to a *BizFail*, so a 7d deadline extension is granted to the buyer, and the right to cancel is granted to the seller. The buyer succeeds in his second attempt (*Pay Success*) while the seller decides not to cancel. In the third scenario, the payment fails three times (a *TecFail* followed by two *BizFail*) without cancellation from the seller, so the business transaction is stopped at *failure count=3 (R8)*. In the last scenario the payment succeeds in the second attempt (*Pay Success*) while the seller successfully exercises his right to cancel (*Seller Canc. Success*) after the buyer's first attempt to pay fails (*Pay BizFail*); if the executions of *Pay Success* and *Seller Canc. Success* conversations overlap, it is possible that (as shown in the figure) the event *Pay Success* is processed at the CCC after *Seller Canc. Success*; consequently, the seller executes a *Refund* conversation that succeeds.

Fig. 2. Execution of payment conversations with success and failure outcomes

5   Conclusions

In this paper we have presented a notation constituting the core of a language to specify e–contracts. A detailed discussion of how our notation meets the requirements stated in Section 1 and of literature relevant to our work is presented in [1]. A notable feature of our work is that we take into account the distributed nature of the underlying computations by paying due attention to timing and message validity constraints as well as the impact of exceptions/failures encountered during business interactions. Work is ongoing on a compiler to translate a rule base written in our notation into one for the rule engine Drools [8]. While in this paper we concentrated our attention on contract description, and indirectly on verification of compliance, contract enforcement is also possible with additions to the core established here, such as an enforcer component to empower the CCC to take a proactive stance during a business partnership.

Acknowledgements

This work has been funded in part by UK EPSRC Platform Grant EP/D037743/1.

References

1. Strano, M., Molina-Jimenez, C., Shrivastava, S.: A rule-based notation to specify executable electronic contracts. Technical Report, School of Computing Science, Newcastle University (2008)
2. Molina-Jimenez, C., Shrivastava, S., Warne, J.: A method for specifying contract mediated interactions. In: Proc. of 9th IEEE Int'l Enterprise Distributed Object Computing Conference (EDOC 2005), pp. 106–115 (2005)
3. RosettaNet: Implementation framework – core specification, http://www.rosettanet.org/
4. ebXML: Business Process Spec. Schema Tech. Spec. v2.0.4 (2006), http://docs.oasis-open.org/ebxml-bp/2.0.4/OS/spec/ebxmlbp-v2.0.4-Spec-os-en.pdf
5. Molina-Jimenez, C., Shrivastava, S., Cook, N.: Implementing Business Conversations with Consistency Guarantees Using Message-Oriented Middleware. In: 11th IEEE Int'l Enterprise Distrib. Object Computing Conf. (EDOC 2007) (2007)
6. Governatori, G., Rotolo, A.: Logic of violations: A Gentzen system for reasoning with contrary–to–duty obligations. Australasian Journal of Logic 4, 193–215 (2006)
7. Strano, M., Molina-Jimenez, C., Shrivastava, S.: A model for checking contractual compliance of business operations. Technical Report N. 1094, School of Computing Science, Newcastle University (2008)
8. JBoss: Drools, http://www.jboss.org/drools/

On Extending RuleML for Modal Defeasible Logic

Duy Hoang Pham[1,2], Guido Governatori[1], Simon Raboczi[2],
Andrew Newman[2], and Subhasis Thakur[2]

[1] National ICT Australia, Queensland Research Laboratory, Brisbane, Australia
[2] School of Information Technology and Electrical Engineering
The University of Queensland, Brisbane, Australia

Abstract. In this paper we present a general methodology to extend Defeasible Logic with modal operators. We motivate the reasons for this type of extension and we argue that the extension will allow for a robust knowledge framework in different application areas. The paper presents an extension of RuleML to capture Modal Defeasible Logic.

1 Introduction

Relations among organizations are guided by sets of rules or policies. A policy can define the privacy requirements of an user, access permissions for a resource, rights of an individual and so on. Many languages have been proposed to write policies. A few examples of these languages are P3P, XACML, SAML. These languages are XML based and use different tags to represent different information to be used in the description of a policy. The growth of the number of these languages and important, and the similarity of concepts these are trying to capture has recently led the W3C to create a special interest group on policy language [22] with the aim of providing a unifying approach to the representation of policies on the web.

A policy can be understood as a set of rules, and the purpose of policy languages (and rule languages in general) is to provide a medium to allow different stakeholders to achieve interoperability by exchanging their (relevant) policies. While the ability to exchange rules is very important, the real key issue is the ability to use and reason with rules in the same way. It might be possible that for some reasons the parties involved in an exchange or rules do not want to adopt the reasoning mechanism of their counterparts. However, they have to realise and understand how the counterparts are going to use the rules, and to consider this in their decision processes.

Rules and proofs are now part of the grand design of the Semantic Web. It has been recognised that the logic part –mainly understood as the OWL family and (fragment) of first order logic– has to be supplemented by rules. Thus the first problem we have to face is to combine logics for reasoning with rules and logics for reasoning with ontologies [7,8,14,23]. The second problem is that while there is only one classical first-order logic but there are many logics for reasoning with rules, and often these logics reflect different and sometimes incompatible facets of reasoning with rules. In addition, we are going to add modal operators and as we will argue in Section 2 even for the same interpretation of a modal operator different logical properties have been proposed. Thus we

N. Bassiliades, G. Governatori, and A. Paschke (Eds.): RuleML 2008, LNCS 5321, pp. 89–103, 2008.

believe that if one wants to be able to share rules with others, it is of paramount importance to be able to specify how to give meaning to the rules and the (modal) operators used in the rules, to enable users to process the information present in the rules in the same way.

The contribution of the paper is manifold. First we will argue that extending rule languages with modal operators offers a very powerful and rich environment to deal with situations where multiple parties are involved and intensional notions are required (Section 2). Deploying any reasoning mechanism for the Web faces an additional challenge: it has to have good computational properties. We defend and motivate our choices against this requirement in Sections 3 and 4. In Section 6.2 we will argue that a rule language should describe the elements of the language but in situations where there are many logics sharing the language, the rule language should provide facilities to describe the logic to be used to process the rules. Here we show how to extend RuleML to capture the descriptions charactersing the extensions with modal operators indentified in Sections 5.1 and 5.2. In Section 7 we outline the implementation of the framework.

2   Modal Logics vs Modalities

Modal logic has been heavily used as a conceptual tool for establishing the foundations of the analysis of epistemic and doxastic notions (i.e., knowledge and belief) in terms of modal operators, paving thus the way to the field of agents and multi-agent systems. In this fields modal operators proved to be very powerful conceptual tools to describe the internal (mental) states of agents as well as interactions among agents. Deontic Logic is the modal logic where the modal operators are interpreted as is nowadays one of the most promising instruments for the formalisation of institutionalised organisation and the mutual relationships (normative position) among the actors in such models. Deontic Logic plays an important role in the formalisation of contracts [9,18].

What we want to stress out here is that modal logic is appropriate to provide a conceptual model for describing agents as well as many other intensional notions, in particular normative notions such as obligations, permissions, rights and so on which are important for policies, e-commerce and e-contract. Given this, the aim of this paper is to provide a computationally oriented non-monotonic rule based account of modal logic for the use and exchange of rules on the Web.

A modal operator qualifies the truth of the expressions it operates on, and many interpretations are possible for modal operator. Given the multiplicity of interpretations and the many facets of modalities, it is not possible to have a one size fits all (or most) situation. In general, there is no single modal logic even for a particular interpretation, and thus the designer of a particular application has to choose case by case which proprieties/principles are satisfied by the modal operators. The designer has to identify which notions are better modelled by modal operators and which are suitable to be captured by predicates.

Given the issues above, a supporter of modalities (particular *ad hoc* predicates whose interpretation is that of modal operators) might argue that modalities offer a more convenient approach since there is no need to create a new logic every time we have a new notion. Everything can be represented in first-order logic. After all, it is hard to distinguish between notions to be modelled by ordinary predicates and notions to be modelled by modal operators. In addition, from a computational point of view first-order logic is semi-decidable while often modal logics are decidable, and there are examples where properties can be encoded easily in modal logic but they require high-order logic representations.

A first answer to this objection is that rather than adding ad hoc predicates to the language, improvements must be made by adding modal operators so as to achieve a richer language that can represent the behaviour of modal notions in a more natural and applicable manner. The advantage of this approach is to incorporate general and flexible reasoning mechanisms within the inferential engine.

A formal representation language should offer concepts close to the notions the language is designed to capture. For example, contracts typically contain provisions about deontic concepts such as obligations, permissions, entitlements, violations and other (mutual) normative positions that the signatories of a contract agree to comply with. Accordingly, a contract language should cater for those notions. In addition, the language should be supplemented by either a formal semantics or facilities to reason with and about the symbols of the language to give meaning to them. As usual, the symbols of the language can be partitioned in two classes: logical symbols and extra logical symbols. The logical symbols are meant to represent general concepts and structures common to every contract while extra logical symbols encode the specific subject matter of given contracts. In this perspective the notions of obligation and permission will be represented by deontic modalities while concepts such as price, service and so on are better captured by predicates since their meaning varies from contract to contract.

In general, we believe that the approach with modal operators is superior to the use of ad hoc predicates at least for the following aspects[1]:

– *Ease of expression and comprehension.* In the modal approach the relationships among modal notions are encoded in the logic and reasoning mechanism while for ad hoc predicates knowledge bases are cluttered with rules describing the logical relationships among different modes/representations of one and the same concept. For example, in a set of rules meant to describe a contract, given the predicate $pay(X)$, we have to create predicates such as $obligatory\_pay(X)$, $permitted\_pay(X)$, ... and rules such as $obligatory\_pay(X) \rightarrow permitted\_pay(X)$ and so on. Thus ad hoc predicates do not allow users to focus only and exclusively on aspects related to the content of a contract, without having to deal with any aspects related to its implementation.
– *Clear and intuitive semantics.* It is possible to give a precise, unambiguous, intuitive and general semantics to the notions involved while each ad hoc predicate requires its own individual interpretation, and in some cases complex constructions (for example reification) are needed to interpret some ad hoc predicates.
– *Modularity.* A current line of research proposes that the combination of deontic operators with operators for speech acts and actions faithfully represent complex

---

[1] In addition to the aspects we discuss here, we would like to point out that it has been argued [13,15] that deontic logic is better than a predicate based representation of obligations and permissions when the possibility of norm violation is kept open. A logic of violation is essential for the representation of contracts where rules about violations are frequent [9].

normative positions such as delegation, empowerment as well as many others that may appear in contracts [16]. In the modal approach those aspects can be added or decomposed modularly without forcing the user to rewrite the predicates and rules to accommodate the new facilities, or to reason at different granularity.

3   Defeasible Logic

Defeasible Logic (DL) [1,20] is a simple, efficient but flexible non-monotonic formalism that can deal with many different intuitions of non-monotonic reasoning [2], and efficient and powerful implementations have been proposed [4,19]. In the last few years the logic and its variants have been applied in many fields.

Knowledge in DL can be represented in two ways: facts and rules.

Facts are indisputable statements, represented either in form of states of affairs (literal and modal literal) and actions that have been performed. Facts are represented by predicates. For example, "the price of the spam filter is $50" is represented by

$$Price(SpamFilter, 50).$$

A *rule*, on the other hand, describes the relationship between a set of literals (premises) and a literal (conclusion), and we can specify how strong the relationship is and the mode the rule connects the antecedent and the conclusion. As usual, rules allow us to derive new conclusions given a set of premises. Since rules have a mode, the conclusions will be modal literals. As far as the strength of rules is concerned we distinguish between *strict rules*, *defeasible rules* and *defeaters*; for the mode we have one set of rules (base rules) describing the inference principles of the basic logic plus one mode for each modal operator of the language (modal rules). As we will see, the idea of modal rules is to introduce modalised conclusions. Accordingly, if we have a modal rule for p for a modal operator $\Box_i$, this means that the rule allows for the derivation of $\Box_i p$.

Strict rules, defeasible rules and defeaters are represented, respectively, by expressions of the form $A_1, \ldots, A_n \rightarrow B$, $A_1, \ldots, A_n \Rightarrow B$ and $A_1, \ldots, A_n \rightsquigarrow B$, where $A_1, \ldots, A_n$ is a possibly empty set of prerequisites and B is the conclusion of the rule. We only consider rules that are essentially propositional. Rules containing free variables are interpreted as the set of their ground instances.

Strict rules are rules in the classical sense: whenever the premises are indisputable then so is the conclusion. Thus, they can be used for definitional clauses. An example of a strict rule is "A 'Premium Customer' is a customer who has spent $10000 on goods":

$$TotalExpense(X, 10000) \rightarrow PremiumCustomer(X).$$

Defeasible rules are rules that can be defeated by contrary evidence. An example of such a rule is "Premium Customer are entitled to a 5% discount":

$$PremiumCustomer(X) \Rightarrow Discount(X).$$

The idea is that if we know that someone is a Premium Customer then we may conclude that she is entitled to a discount *unless there is other evidence suggesting that she may not be* (for example if she buys a good in promotion).

Defeaters are a special kind of rules. They are used to prevent conclusions not to support them. For example:

$$SpecialOrder(X), PremiumCustomer(X) \rightsquigarrow \neg Surcharge(X).$$

This rule states that premium customers placing special orders might be exempt from the special order surcharge. This rule can prevent the derivation of a "surcharge" conclusion. However, it cannot be used to support a "not surcharge" conclusion.

DL is a "skeptical" non-monotonic logic, meaning that it does not support contradictory conclusions.[2] Instead, DL seeks to resolve conflicts. In cases where there is some support for concluding A but also support for concluding $\neg A$, DL does not conclude neither of them (thus the name "skeptical"). If the support for A has priority over the support for $\neg A$ then A is concluded.

As we have alluded to above, no conclusion can be drawn from conflicting rules in DL unless these rules are prioritised. The *superiority relation* is used to define priorities among rules, that is, where one rule may override the conclusion of another rule. For example, given the defeasible rules

$$r : PremiumCustomer(X) \Rightarrow Discount(X)$$
$$r' : SpecialOrder(X) \Rightarrow \neg Discount(X)$$

which contradict one another, no conclusive decision can be made about whether a Premium Customer, who has placed a special order, is entitled to the 5% discount. But if we introduce a superiority relation $>$ with $r' > r$, we can indeed conclude that special orders are not subject to discount.

We now give a short informal presentation of how conclusions are drawn in DL. Let D be a theory in DL (i.e., a collection of facts, rules and a superiority relation). A *conclusion* of D is a tagged literal and can have one of the following four forms:

$+\Delta q$ meaning that q is definitely provable in D (i.e., using only facts and strict rules).
$-\Delta q$ meaning that we have proved that q is not definitely provable in D.
$+\partial q$ meaning that q is defeasibly provable in D.
$-\partial q$ meaning that we have proved that q is not defeasibly provable in D.

Strict derivations are obtained by forward chaining of strict rules while a defeasible conclusion p can be derived if there is a rule whose conclusion is p, whose prerequisites (antecedent) have either already been proved or given in the case at hand (i.e. facts), and any stronger rule whose conclusion is $\neg p$ has prerequisites that fail to be derived. In other words, a conclusion p is derivable when:

- p is a fact; or
- there is an applicable strict or defeasible rule for p, and either
  - all the rules for $\neg p$ are discarded (i.e., are proved to be not applicable) or
  - every applicable rule for $\neg p$ is weaker than an applicable strict[3] or defeasible rule for p.

The formal definitions of derivations in DL are in the next section.

---

[2] To be precise contradictions can be obtained from the monotonic part of a defeasible theory, i.e., from facts and strict rules.

[3] Notice that a strict rule can be defeated only when its antecedent is defeasibly provable.

4 Modal Defeasible Logic

As we have seen in Section 1, modal logics have been put forward to capture many different notions somehow related to the intensional nature of agency as well as many other notions. Usually modal logics are extensions of classical propositional logic with some intensional operators. Thus, any modal logic should account for two components: (1) the underlying logical structure of the propositional base and (2) the logic behaviour of the modal operators. Alas, as is well-known, classical propositional logic is not well suited to deal with real life scenarios. The main reason is that the descriptions of real-life cases are, very often, partial and somewhat unreliable. In such circumstances, classical propositional logic might produce counterintuitive results insofar as it requires complete, consistent and reliable information. Hence any modal logic based on classical propositional logic is doomed to suffer from the same problems.

On the other hand, the logic should specify how modalities can be introduced and manipulated. Some common rules for modalities are, e.g., Necessitation (from $\vdash \phi$ infer $\vdash \Box\phi$) and RM (from $\vdash \phi \rightarrow \psi$ infer $\vdash \Box\phi \rightarrow \Box\psi$). Both dictates conditions to introduce modalities purely based on the derivability and structure of the antecedent. These rules are related to the well-known problem of logical omniscience and put unrealistic assumptions on the capability of an agent. However, if we take a constructive interpretation, we have that if an agent can build a derivation of φ then she can build a derivation of $\Box\varphi$. We want to maintain this intuition here, but we want to replace derivability in classical logic with a practical and feasible notion like derivability in DL. Thus, the intuition behind this work is that we are allowed to derive $\Box_i p$ if we can prove p with the mode $\Box_i$ in DL.

To extend DL with modal operators we have two options: 1) to use the same inferential mechanism as basic DL and to represent explicitly the modal operators in the conclusion of rules [21]; 2) introduce new types of rules for the modal operators to differentiate between modal and factual rules.

For example, the "deontic" statement "The Purchaser shall follow the Supplier price lists" can be represented as

$$AdvertisedPrice(X) \Rightarrow O_{purchaser}Pay(X)$$

if we follow the first option and

$$AdvertisedPrice(X) \Rightarrow_{O_{purchaser}} Pay(X)$$

according to the second option, where $\Rightarrow_{O_{purchaser}}$ denotes a new type of defeasible rule relative to the modal operator $O_{purchaser}$. Here, $O_{purchaser}$ is the deontic "obligation" operator parametrised to an actor/role/agent, in this case the purchaser.

The differences between the two approaches, besides the fact that in the first approach there is only one type of rules while the second accounts for factual and modal rules, is that the first approach has to introduce the definition of p-incompatible literals (i.e., a set of literals that cannot be hold when p holds) for every literal p. For example, we can have a modal logic where $\Box p$ and $\neg p$ cannot be both true at the same time. Moreover, the first approach is less flexible than the second: in particular in some cases it must account for rules to derive $\Diamond p$ from $\Box p$; similarly conversions (see Section 5.2) require additional operational rules in a theory, thus the second approach seems to offer a more conceptual tool than the first one. The second approach can use different proof conditions based on the modal rules to offer a more fine grained control over the modal operators and it allows for interaction between modal operators.

As usual with non-monotonic reasoning, we have to specify 1) how to represent a knowledge base and 2) the inference mechanism used to reason with the knowledge base. The language of Modal Defeasible Logic consists of a finite set of modal operators $Mod = \{\Box_1,\ldots,\Box_n\}$ and a (numerable) set of atomic propositions $Prop = \{p,q,\ldots\}$.[4]

We supplement the usual definition of literal (an atomic proposition or the negation of it), with the following clauses

- if l is a literal then $\Box_i l$, and $\neg\Box_i l$, are literals if l is different from $\Box_i m$, and $\neg\Box_i m$, for some literal m.

The above condition prevents us from having sequences of modalities where we have successive occurrences of one and the same modality; however, iterations like $\Box_i\Box_j$ and $\Box_i\Box_j\Box_i$ are legal in the language.

Given a literal l with $\sim l$ we denote the complement of l, that is, if l is a positive literal p then $\sim l = \neg p$, and if $l = \neg p$ then $\sim l = p$.

According to the previous discussion a Modal Defeasible Theory D is a structure $(F,R,\succ)$ where F is a set of facts (literals or modal literals), $R = R^B \cup \bigcup_{1\leq i\leq n} R^{\Box_i}$, where R^B is the set of base (un-modalised) rules, and each $R^{\Box_i}$ is the set of rules for $\Box_i$ and $\succ\subseteq R \times R$ is the superiority relation. A rule r is an expression $A(r) \hookrightarrow_X C(r)$ such that ($\hookrightarrow\in \{\rightarrow,\Rightarrow,\rightsquigarrow\}$, X is B, for a base rule, and a modal operator otherwise), $A(r)$ the antecedent or body of r is a (possible empty) set of literals and modal literals, and $C(r)$, the consequent or head of r is a literal if r is a base rule and either a literal or a modal literal Yl where Y is a modal operator different from X. Given a set of rules R we use R_{sd} to denote the set of strict and defeasible rules in R, and $R[q]$ for the set of rules in R whose head is q.

The derivation tags are now indexed with modal operators. Let X range over Mod. A conclusion can now have the following forms:

$+\Delta_X q$: q is definitely provable with mode X in D (i.e., using only facts and strict rules of mode X).

$-\Delta_X q$:   we have proved that q is not definitely provable with mode X in D.

$+\partial_X q$: q is defeasibly provable with mode X in D.

$-\partial_X q$:   we have proved that q is not defeasibly provable with mode X in D.

Then if we can prove $+\partial_{\Box_i} q$, then we can assert $\Box_i q$.

Formally provability is based on the concept of a *derivation* (or proof) in D. A derivation is a finite sequence $P = (P(1),\ldots,P(n))$ of tagged literals satisfying the proof conditions (which correspond to inference rules for each of the kinds of conclusion). $P(1..n)$ denotes the initial part of the sequence P of length n.

---

[4] The language can be extended to deal with other notions. For example to model agents, we have to include a (finite) set of agents, and then the modal operators can be parameterised with the agents. For a logic of action or planning, it might be appropriate to add a set of atomic actions/plans, and so on depending on the intended applications.

Before introducing the proof conditions for the proof tags relevant to this paper we provide some auxiliary notions.

Let # be either Δ or ∂. Given a proof $P = (P(1),\ldots,P(n))$ in D and a literal q we will say that q is Δ-*provable* in P, or simply Δ-provable, if there is a line $P(m)$ of the derivation such that either:

1. if $q = l$ then
   - $P(m) = +\#l$ or
   - $\Box_i l$ is #-provable in $P(1..m-1)$ and $\Box_i$ is reflexive[5]
2. if $q = \Box_i l$ then
   - $P(m) = +\#_i l$ or
   - $\Box_j \Box_i l$ is #-provable in $P(1..m-1)$, for some $j \neq i$ such that $\Box_j$ is reflexive.
3. if $q = \neg \Box_i l$ then
   - $P(m) = -\#_i l$ or
   - $\Box_j \neg \Box_i l$ is #-provable in $P(1..m-1)$, for some $j \neq i$ such that $\Box_j$ is reflexive.

In a similar way we can define a literal to be Δ- and ∂-rejected by taking, respectively, the definition of Δ-provable and ∂-provable and changing all positive proof tags into negative proof tags, adding a negation in front of the literal when the literal is prefixed by a modal operator $\Box_j$, and replacing all the *or*s by *and*s. Thus, for example, we can say that a literal $\Box_i l$ is ∂-rejected if, in a derivation, we have a line $-\partial_i l$, and the literal $\neg \Box_i \neg l$ is ∂-rejected if we have $+\partial_i \neg l$ and so on.

Let X be a modal operator and # is either Δ or ∂. A literal l is $\#_X$-*provable* if the modal literal Xl is #-provable; l is $\#_X$-rejected if the literal Xl is #-rejected.

Based on the above definition of provable and rejected literals we can give the conditions to determine whether a rule is applicable or the rule cannot be used to derive a conclusion (i.e., the rule is discarded).

The proof conditions for $+\Delta$ correspond to monotonic forward chaining of derivations and, for space limitations are not given here (see [1,10] for the definitions).

Let X be a modal operator or B. Given a rule r we will say that the rule is ∂_X-*applicable* iff

1. $r \in R^X$ and $\forall a_k \in A(r)$, a_k is ∂-provable; or
2. if $X \neq B$ and $r \in R^B$, i.e., r is a base rule, then $\forall a_k$, a_k is ∂_X-provable.

Given a rule r we will say that the rule is ∂_X-*discarded* iff

1. $r \in R^X$ and $\exists a_k \in A(r)$, a_k is ∂-rejected; or
2. if $X \neq B$ and $r \in R^B$, i.e., r is a base rule, then $\exists a_k$, a_k is ∂_X-rejected.

We give now the proof condition for defeasible conclusions (i.e., conclusions whose tag is $+\partial$). Defeasible derivations have an argumentation like structure divided in three phases. In the first phase, we put forward a supported reason (rule) for the conclusion we want to prove. Then in the second phase, we consider all possible (actual and not) reasons against the desired conclusion. Finally, in the last phase, we have to rebut all

---

[5] A modal operator $\Box_i$ is reflexive iff the truth of $\Box_i \phi$ implies the truth of ϕ. In other words $\Box_i$ is reflexive when we have the modal axiom $\Box_i \phi \rightarrow \phi$.

the counterarguments. This can be done in two ways: we can show that some of the premises of a counterargument do not obtain, or we can show that the argument is weaker than an argument in favour of the conclusion. This is formalised by the following (constructive) proof conditions.

$+\partial_X$: If $P(n+1) = +\partial_X q$ then
    1) $+\Delta_X q \in P(1..n)$, or
    2) $-\Delta_X \sim q \in P(1..n)$ and
        2.1) $\exists r \in R_{sd}[q]$: r is ∂_X-applicable and
        2.2) $\forall s \in R[\sim q]$ either s is ∂_X-discarded or
             $\exists w \in R[q]$: w is ∂_X-applicable and $w \succ s$.

The above condition is, essentially, the usual condition for defeasible derivations in DL, we refer the reader to [1,10,20] for more thorough treatments. The only point we want to highlight here is that base rules can play the role of modal rules when all the literals in the body are $\partial_{\Box_i}$-derivable. Thus, from a base rule $a, b \Rightarrow_B c$ we can derive $+\partial_{\Box_i} c$ if both $+\partial_{\Box_i} a$ and $+\partial_{\Box_i} b$ are derivable while this is not possible using the rule $a, \Box_i b \Rightarrow_B c$ (see Section 5.2).

5   Modal Defeasible Logic with Interactions

Notice that the proof condition for $+\partial$ given in Section 3 and then those for the other proof tags are the same as those of basic DL as given in [1]. What we have done is essentially to consider $n+1$ non-monotonic consequence relation defined in DL and compute them in parallel. In the previous sections, we have argued that one of the advantages of modal logic is the ability to deal with complex notions composed by several modalities, or by interactions of modal operators. Thus, we have to provide facilities to represent such interactions. In Modal DL it is possible to distinguish two types of interactions: conflicts and conversions. In the next two sections, we will motivate them and we show how to capture them in our framework.

5.1   Conflicts

Let us take a simple inclusion axiom of multi-modal logic relating two modal operators $\Box_1$ and $\Box_2$: $\Box_1 \phi \rightarrow \Box_2 \phi$. The meaning of this axiom is that every time we are able to prove $\Box_1 \phi$, then we are able to prove $\Box_2 \phi$. Thus, given the intended reading of the modal operators in our approach –a modal operator characterises a derivation using a particular mode, it enables us to transform a derivation of $\Box_1 \phi$ into a derivation of $\Box_2 \phi$. If the logic is consistent, we also have that $\Box_1 \phi \rightarrow \Box_2 \phi$ implies that it is not possible to prove $\Box_2 \neg \phi$ given $\Box_1 \phi$, i.e., $\Box_1 \phi \rightarrow \neg \Box_2 \neg \phi$. However, this idea is better illustrated by the classically equivalent formula $\Box_1 \phi \wedge \Box_2 \neg \phi \rightarrow \bot$. When the latter is expressed in form of the inference rule

$$\frac{\Box_1 \phi, \Box_2 \neg \phi}{\bot} \tag{1}$$

it suggests that it is not possible to obtain $\Box_1 \phi$ and $\Box_2 \neg \phi$ together. This does not mean that $\Box_1 \phi$ implies $\Box_2 \phi$, but that the modal operators $\Box_1$ and $\Box_2$ are in conflict with each other. Modal DL is able to differentiate between the two formulations: For the inclusion version (i.e., $\Box_1\phi \rightarrow \Box_2\phi$) what we have to do is just to add the following clause to the proof conditions for $+\partial_{\Box_2}$ (and the other proof tags accordingly) with the condition

$$+\partial_{\Box_1} q \in P(1..n)$$

For the second case (i.e., $\Box_1\phi \land \neg\Box_2\phi \rightarrow \bot$), we have to give a preliminary definition.

Given a modal operator $\Box_i$, $\mathscr{F}(\Box_i)$ is the set of modal operators in conflict with $\Box_i$. If the only conflict axiom we have is $\Box_1\phi \land \Box_2\phi \rightarrow \bot$ then $\mathscr{F}(\Box_1) = \{\Box_2\}$. With $R^{\mathscr{F}(\Box_i)}$ we denote the union of rules in all $R^{\Box_j}$ where $\Box_j \in \mathscr{F}(\Box_i)$. At this point to implement the proof condition for the conflict all we have to do is to replace clause 2.2 of the definition of $+\partial_{\Box_i} q$ with the clause

2.2)$\forall s \in R^{\mathscr{F}(\Box_i)}[\sim q]$ either s is ∂_X-discarded or
$\exists w \in R[q]$: w is ∂_X-applicable and $w \succ s$.

The notion of conflict has been proved useful in the area of cognitive agents, i.e., agent whose rational behaviour is described in terms of mental and motivational attitudes including beliefs, intentions, desires and obligations. Classically, agent types are characterised by stating conflict resolution methods in terms of orders of overruling between rules [6,10]. For example, an agent is *realistic* when rules for beliefs override all other components; she is *social* when obligations are stronger than the other components with the exception of beliefs. Agent types can be characterised by stating that, for any types of rules X and Y, for every r and r', $r \in R^X[q]$ and $r' \in R^Y[\sim q]$, we have that $r > r'$.

5.2 Conversions

Another interesting feature that could be explained using our formalism is that of *rule conversion*. Indeed, this feature allows us to model the interactions between different modal operators. In general, notice that in many formalisms it is possible to convert from one type of conclusion into a different one. For example, the right weakening rule of non-monotonic consequence relations (see [17])

$$\frac{B \vdash C \quad A \mathrel{\mid\!\sim} B}{A \mathrel{\mid\!\sim} C}$$

allows the combination of non-monotonic and classical consequences.

Suppose that a rule of a specific type is given and all the literals in the antecedent of the rule are provable in one and the same modality. If so, is it possible to argue that the conclusion of the rule inherits the modality of the antecedent? To give an example, suppose we have that $p, q \Rightarrow_{\Box_i} r$ and that we obtain $+\partial_{\Box_j} p$ and $+\partial_{\Box_j} q$. Can we conclude $\Box_j r$? In many cases this is a reasonable conclusion to obtain.

For this feature we have to declare which modal operators can be converted and the target of the conversion. Given a modal operator $\Box_i$, with $\mathscr{V}(\Box_i)$ we denote the set of modal operators $\Box_j$ that can be converted to $\Box_i$. In addition, we assume that base rules can be converted to all other types of rules. The condition to have a successful conversion of a rule for $\Box_j$ into a rule for $\Box_i$ is that all literals in the antecedent of the rules are provable modalised with $\Box_i$. Formally we have thus to add (disjunctively) in the support phase (clause 2.1) of the proof condition for $\partial_{\Box_i}$ the following clause

2.1b) $\exists r \in R^{\gamma(\Box_i)}[q]$ such that r is $\partial_{\Box_i}$-applicable

The notion of conversion enables us to define new interesting agent types [10].

We conclude this section with a formalisation of the Yale Shooting Problem that illustrates the notion of conversion. Let INT be the modal operator for intention. The Yale Shooting Problem can be described as follows[6]

$$liveAmmo, load, shoot \Rightarrow_B kill$$

This rule encodes the knowledge of an agent that knows that loading the gun with live ammunitions, and then shooting will kill her friend. This example clearly shows that the qualification of the conclusions depends on the modalities relative to the individual acts "load" and "shoot". In particular, if the agent intends to load and shoot the gun (INT(*load*), INT(*shoot*)), then, since she knows that the consequence of these actions is the death of her friend, she intends to kill him ($+\partial_{\text{INT}}kill$). However, in the case she has the intention to load the gun ($+\partial_{\text{INT}}load$) and for some reason shoot it (*shoot*), then the friend is still alive ($-\partial kill$).

6 RuleML

Starting with the RuleML 0.91 XML Schema for Datalog with classical negation, we extended the syntax to support defeasible rules and modal operators.

6.1 Defeasible Rule Markup

RuleML already supports strict rules via the `Implies` element and allows them to be named using the `oid` element. We need to extend the syntax to express defeasible rules, defeaters, and superiority relations.

To add defeasible rules and defeaters as described in §3, we borrow syntax from the DR-DEVICE rule language [4]. We add a `@ruletype` attribute to the `Implies` element, allowing it to take one of three values: `strictrule`, `defeasiblerule` or `defeater`. Because `strictrule` is implied when `@ruletype` is absent, when non-defeasible RuleML rulesets are imported their rules are correctly considered strict.

DR-DEVICE expresses the superiority relation by using the `@superior` attribute on the superior rule as a link to the `@ruleID` label of the inferior rule. We found this unsuitable because we may need to mark a rule as superior to more than one other rule, and an XML element can only bear a single `@superior` attribute. Using the scheme from [9, §5] instead, we explicitly represent the superiority relation using the distinguished predicate `Override`.

6.2 Modal Operator Markup

In §2 we argued against modality predicates such as those proposed in [5, §4]. Furthermore, in §4 we proposed two alternatives, modal operators and modal rules.

---

[6] Here we will ignore all temporal aspects and we will assume that the sequence of actions is done in the correct order.

To support the first alternative, we introduce a Mode element. The @modetype at-tribute is a URI-valued identifier for the intended semantics of the modal operator, e.g. necessity, belief, obligation. The Mode may optionally contain a single parameters el-ement whose zero or more children are used to further distinguish modes, e.g. between the beliefs of various agents, or between time instants in the case of a temporal operator. Two modes are identical if their @modetype and all their parameters are equal. For example, $r1 : AdvertisedPrice(X) \Rightarrow O_{purchaser}Pay(X)$ is represented as

```
<Implies ruletype="defeasiblerule">
  <oid><Ind>r1</Ind></oid>
  <head>
    <Mode modetype="http://www.example.org/obligation">
      <parameters>
        <Ind>purchaser</Ind>
      </parameters>
      <Atom><Rel>Pay</Rel><Var>X</Var></Atom>
    </Mode>
  </head>
  <body>
    <Atom><Rel>AdvertisedPrice</Rel><Var>X</Var></Atom>
  </body>
</Implies>
```

To support the second alternative, modal rules, we introduce a mode element which may appear as a child of the Implies element. It requires the same @modetype attribute as the Mode element. Its zero or more children distinguish the mode in the same way as the children of a Mode's parameters. For example, $r2 : AdvertisedPrice(X) \Rightarrow O_{purchaser} Pay(X)$ is represented as

```
<Implies ruletype="defeasiblerule">
  <oid><Ind>r2</Ind></oid>
  <mode modetype="http://www.example.org/obligation">
    <Ind>purchaser</Ind>
  </mode>
  <head>
    <Atom><Rel>Pay</Rel><Var>X</Var></Atom>
  </head>
  <body>
    <Atom><Rel>AdvertisedPrice</Rel><Var>X</Var></Atom>
  </body>
</Implies>
```

6.3 Modal Interactions

The conflict and conversion interactions introduced in §5 are not represented in RuleML. Instead, we express them in a separate document with its own custom XML Schema. This is an additional input file used to configure the reasoner. It lists the supported modes, identifying them globally using the same URIs referenced by the @modetype attributes in the rules, and locally to the document with short XML IDs. These IDs are then used to succinctly list any conflict sets and conversion pairs. The following shows an example configuration file for a 'social' agent [10], that is an agent whose obligations prevail over her intentions and beliefs can be used to derive non primitive intentions and obligations.

```xml
<?xml version="1.0" encoding="UTF-8"?>
<ModeSet xmlns="http://www.example.org/modeset-ns"
    xmlns:ruleml="http://www.ruleml.org/0.91/xsd"
    xmlns:xs="http://www.w3.org/2001/XMLSchema"
    xmlns:xsi="http://www.w3.org/2001/XMLSchema-instance"
    xsi:schemaLocation="http://www.example.org/xsd/ruleset.xsd">
  <Mode id="BEL1" href="http://www.example.org/mode/belief">
    <ruleml:Ind>agent1</ruleml:Ind>
  </Mode>
  <Mode id="OBL" href="http://www.example.org/mode/obligation"/>
  <Mode id="INT1" href="http://www.example.org/mode/intention">
    <ruleml:Ind>agent1</ruleml:Ind>
  </Mode>
  <Conflict between="OBL INT1"/>
  <Conversion from="BEL1" to="INT1"/>
  <Conversion from="BEL1" to="OBL"/>
</ModeSet>
```

7   Implementation

The reasoning process of Modal DL has three phases. In the pre-processing phase, the theory in the RuleML format are loaded into the mechanism and is transformed into an equivalent theory without superiority relation and defeaters. In the next phase, the rule loader, which parses the theory obtained in the first phase, generates the data structure for the inferential phase. Finally, the inference engine applies modifications to the data structure, where at every step it reduces the complexity of the data structure.

Theory transformation: The transformation operates in three steps. The first two steps remove the defeaters rules and the superiority relation among rules by applying the transformations similar to those of [10]. Essentially, the hierarchy of the modal operators is generated from the conflicting relationship among these operators. The modal operator on the top of the hierarchy plays the role of the *BEL* operator as in [10]. This amounts to take the rules for the modal operator at the top of the hierarchy as the set of base rules. The third step performs conversions of every modal rule into a rule with a new modal operator as specified by the theory.

Rule loader: The rule loader creates a data structure as follows: for every (modal) literal in the theory, we create an entry whose structure includes:

- a list of (pointers to) rules having the literal in the head. In order to simplify the data structure, a modal literal from the head of a rule is built from the head atom and the modal operator of the corresponding rule.
- a list of (pointers to) rules having the literal in the body

- a list of (pointers to) entries of complements of the literal. Notice that the complements of a literal should take into account of the occurrence the modal operator. For example, the complements of the literal $\Box_i l$ are $\neg \Box_i l$ and $\Box_i \sim l$; if the operator is reflexive we have to include also l as a complement of $\Box_i l$.
- a list of entries of literals which conflict with the literal. The conflict relationship is derived from the conflicting modal operators dictated by the theory. In addition, a modal literal $\Box_i l$ always conflicts with $\sim l$ when $\Box_i$ is reflexive.

In order to improve the computational performance, every list in the data structure is implemented as a hash table.

Inferential engine: The Engine is based on an extension of the Delores algorithm proposed in [19] as a computational model of Basic Defeasible Logic. In turn, the engine

- Assert each fact (as a literal) as a conclusion and removes the literal from the rules, where the literal positively occurs in the body, and "deactivate" the rules where either its complements or its conflicting literals occur in the body.
- Scan the list of active rules for rules with the empty body. Take the (modal) literal from the head, remove the rule, and put the literal into the pending facts. The literal is removed from the pending facts and adds to the list of facts if either there is no such rule (of the appropriate type) whose head contains the complements of the literal or literals with conflicting modes, or it is impossible to prove these literals.
- It repeats the first step.
- The algorithm terminates when one of the two steps fails.[7] On termination, the algorithm outputs the set of conclusions from the list of facts in the RuleML format.

8 Conclusion

To sum up the contribution of the paper is manyfold. We have argued that rule languages for the Semantic Web can benefit from modal extensions. However, given the multiplicity of interpretations of modal operators (as well as the different intuition behind execution model of rule systems) present a further challenge. An interchange language should be able to provide not only the syntax to represent rule, but it should provide facilities to describe how the rules should be processed (i.e., what the is the logic to be used to interpret the rules). On this respect we have identified the basic mechanisms to relate modal operators in a rule language (conflict and conversion).

The framework we have outlined in the previous sections has proven robust enough to represent and reason with different scenarios and applications, from business contracts [9] to normative reasoning [12], policy based cognitive agents [10] and workflow systems [11]. The main reason of the success, we believe, is due to the fact that Modal DL conceptually strengthen the expressive power of DL with modal operators, but at the same time it maintains the constructive and computational flavour of DL. Indeed, we have proved that the complexity of Modal DL as outlined here is linear [10]. This makes the logic very attractive from the knowledge representation point of view.

---

[7] This algorithm outputs $+\partial$; $-\partial$ can be computed by an algorithm similar to this with the "dual actions". For $+\Delta$ we have just to consider similar constructions where we examine only the first parts of step 1 and 2. $-\Delta$ follows from $+\Delta$ by taking the dual actions.

References

1. Antoniou, G., Billington, D., Governatori, G., Maher, M.J.: Representation results for defeasible logic. ACM Transactions on Computational Logic 2(2), 255–287 (2001)
2. Antoniou, G., Billington, D., Governatori, G., Maher, M.J., Rock, A.: A family of defeasible reasoning logics and its implementation. In: Proc. ECAI 2000, pp. 459–463. IOS Press, Amsterdam (2000)
3. Antoniou, G., Boley, H. (eds.): RuleML 2004. LNCS, vol. 3323. Springer, Heidelberg (2004)
4. Bassiliades, N., Antoniou, G., Vlahavas, I.P.: A defeasible logic reasoner for the semantic web. International Journal on Semantic Web and Information Systems 1(2), 1–41 (2006)
5. Boley, H.: The RuleML family of web rule languages. In: Proc 4th PPSWR, pp. 1–17. Springer, Heidelberg (2006)
6. Broersen, J., Dastani, M., Hulstijn, J., van der Torre, L.: Goal generation in the BOID architecture. Cognitive Science Quarterly 2(3-4), 428–447 (2002)
7. Eiter, T., Lukasiewicz, T., Schindlauer, R., Tompits, H.: Well-founded semantics for description logic programs in the semantic web. In: [3], pp. 81–97
8. Governatori, G.: Defeasible description logics. In: [3], pp. 98–112
9. Governatori, G.: Representing business contracts in RuleML. International Journal of Cooperative Information Systems 14(2-3), 181–216 (2005)
10. Governatori, G., Rotolo, A.: BIO Logical Agents: Norms, Beliefs, Intentions in Defeasible Logic. Journal of Autonomous Agents and Multi-Agents (2008)
11. Governatori, G., Rotolo, A., Sadiq, S.: A model of dynamic resource allocation in workflow systems. In: Database Technology 2004, vol. CRPIT 27, pp. 197–206. ACS (2004)
12. Governatori, G., Rotolo, A., Sartor, G.: Temporalised normative positions in defeasible logic. In: Proc. ICAIL 2005, pp. 25–34. ACM Press, New York (2005)
13. Herrestad, H.: Norms and formalization. In: Proc. ICAIL 1991, pp. 175–184. ACM Press, New York (1991)
14. Horrocks, I., Patel-Schneider, P.F., Boley, H., Tabet, S., Grosof, B., Dean, M.: SWRL: A semantic web rule language combining owl and ruleml. W3C Member Submission, 21 May (2004), http://www.w3.org/Submission/SWRL/
15. Jones, A.J.I., Sergot, M.: On the characterization of law and computer systems: the normative systems perspective. In: Deontic logic in computer science: normative system specification, pp. 275–307. John Wiley and Sons Ltd., Chichester (1993)
16. Jones, A.J.I., Sergot, M.: A formal characterisation of institutionalised power. Journal of the IGPL 4(3), 429–445 (1996)
17. Kraus, S., Lehmann, D., Magidor, M.: Nonmonotonic reasoning, preferential models and cumulative logics. Artificial Intelligence 44, 167–207 (1990)
18. Lee, R.M.: A logic model for electronic contracting. Decision Support Systems 4, 27–44 (1988)
19. Maher, M.J., Rock, A., Antoniou, G., Billignton, D., Miller, T.: Efficient defeasible reasoning systems. International Journal of Artificial Intelligence Tools 10(4), 483–501 (2001)
20. Nute, D.: Defeasible logic. In: Handbook of Logic in Artificial Intelligence and Logic Programming, vol. 3, pp. 353–395. Oxford University Press, Oxford (1994)
21. Nute, D.: Norms, priorities and defeasibility. In: Norms, Logics and Information Systems. New Studies in Deontic Logic, pp. 83–100. IOS Press, Amsterdam (1998)
22. Pling – w3c policy languages interest group (Accessed November 1, 2007) (2007), http://www.w3.org/Policy/pling/
23. Wang, K., Billington, D., Blee, J., Antoniou, G.: Combining description logic and defeasible logic for the semantic web. In: [3], pp. 170–181

Adding Uncertainty to a Rete-OO Inference Engine

Davide Sottara[1], Paola Mello[1], and Mark Proctor[2]

[1] DEIS, Facolta di Ingegneria, Universita di Bologna
Viale Risorgimento 2, 40131 Bologna (BO) Italy
[2] JBoss, a division of Red Hat

Abstract. The RETE algorithm has been used to implement first-order logic based inference engines and its object-oriented extension allows to reason directly over entities rather than predicates. One of the limitations of FOL is its inability to deal with uncertainty, although it exists in many forms and it is typical of the way humans reason. In this paper, the steps of a general uncertain reasoning are outlined, without choosing a specific type or representation of uncertainty. Then, the process is translated into a further extension of the RETE networks, showing a possible architecture allowing a Rete-OO based engine to reason with uncertain rules. This architecture is being implemented in the Drools rule engine.

1 Introduction

Rule Based Systems have been widely used, primarily with the role of Expert Systems, because of their high degree of flexibility and understandability. Programming with declarative Knowledge Bases may be easier when defining problems in complex domains and the adopted symbolic languages, quite similar to the natural one, allow an easy interaction with the users, including queries and explanations. Moreover, typical KBs are incremental and robust: rules can be added and removed easily to modify the functionality of the whole system without compromising it, at least as long as the modifications are consistent and local.

Depending on the particular problem, an algorithmic approach may still be more effective (or efficient): in the field of Artificial Intelligence this role is covered by connectionist techniques such as neural networks, genetic algorithms, adaptive classifiers and various others. They, however, lack the clarity and consciousness of a symbolic system and so it is common place to integrate different modules of this type into hybrid systems to have a rule base reason over, control and explain the results of the underlying evaluators. Because of these properties, Rule Based Systems are finding applications in new contexts where coordination and the exchange of information in general is perhaps more important than computation itself, such as the Semantic Web. In order to overcome the heterogeneity of the existing contexts, RuleML [10] has been proposed as a standard yet extensible language for rule definition.

N. Bassiliades, G. Governatori, and A. Paschke (Eds.): RuleML 2008, LNCS 5321, pp. 104–118, 2008.

Rules written in RuleML are guaranteed to have unambiguous semantics and can be exchanged between applications, but are essentially generalized First-Order Logic predicates. The main limitation of FOL is its inability to deal with uncertainty in a native manner, while uncertainty exists in many forms in human reasoning, from stochastic uncertainty (head will appear when a coin is flipped) to vague uncertainty (there are more or less N objects in a box). As will be shown in detail in section 2, many logics have been formalized to model different facets of uncertainty and the languages have been extended accordingly ([15], [5]), trying to provide a unifying framework for expressing the slightly different concepts. However, not all inference engines have been upgraded in parallel, while a single, general purpose engine capable of processing uncertain rules written in one or more standard languages would be preferrable to implementations tailored ad hoc on the individual rule sets.

Suppose, for example, that a system needs to react to the presence of the authors of a certain paper on rule engines: in many cases the only available specification may sound like

IF there is a paper about Drools, authored by a young, maybe tall man called Davide or a person named Paola THEN... .

There may be uncertainty about the identity of these people, their description, or even the topic of the paper: some details may be vague (e.g. young), others may be incomplete or imprecise (e.g. first names instead of family names), other specific but stated in a manner different than the one expected (e.g. Drools is a rule engine?). Uncertainty initially appears in the evaluation of the different features, provided that the concepts to be evaluated have been clearly defined - what does "young"mean? what does "equal"mean? - then must be combined - again, what does "or"mean?

When dealing with rule-based production systems, the RETE algorithm [12] is commonly used to compile Rule Bases into efficient inference engines based on the concept of production system. The goal of this paper is to show that a limited number of modifications, detailed in in section 3, is sufficient to have a RETE network evaluate an uncertain logic program. At the same time, to provide the necessary degree of flexibility, the nodes of the network are given a more abstract definition, allowing the user to plug in custom evaluators and operators to specify the concrete behaviour at runtime.

2   Uncertain Logics

First order logic has been used in the attempt to formalize human reasoning for computational purposes. Even if important practical results have been obtained, it has been debated whether people do reason according to its principles. Many extensions have been developed to capture different facets in expressiveness, several of which have been grouped in the class of uncertain logics.

The term "uncertainty"actually encompasses different concepts. Stochastic uncertainty is perhaps the most common: it arises whenever an actor lacks complete information about the outcome of an event, often because it will take place in an inaccessible location or in the future. This type of uncertainty is measured using probability, which in turn may be estimated using a frequentist approach by repeated trials (e.g. counting the number of heads in a sequence of coin tosses) or by a subjective evaluation in case of unique events (e.g. the probability that it will rain tomorrow).

Probability is different from vagueness, which describes events known or just expressed with some degree of imprecision: it becomes especially useful when the events have a continuous range of possible outcomes, such as the age of a person. In many cases a summary descriptor such as "young"or "old"may be enough, and it should not be forgotten that even saying that they are x years old is an approximation because it ignores months, days and so on.

A third concept is confidence, or certainty: it measures the strength of a statement, whether it is certain or not, either as an absolute value or in comparison to others. It distinguishes solid facts from those lacking enough support and can be used to rule out conflicts or inconsistencies.

Confidence has been the first type of uncertainty implemented in expert systems (MYCIN); after that, probabilistic logic has usually been embedded in Bayesian networks-based systems while fuzzy-logic based systems use vagueness extensively.

Determining which type of uncertainty is more appropriate to model a domain is a very complex and problem-specific task. The issue will not be addressed further in this work; instead it will be assumed that an Expert has provided the uncertain knowledge which is to be formalized and then processed. In general, the number of languages and formalisms is large in proportion to the number of open, general purpose tools and shells capable of processing it. The development of fuzzy RuleML aims at standardizing the existing language in order to facilitate interoperability: the extension proposed in [5] shows that many types of logic can be unified modelling them as concrete instances of operator classes working on instances of different (truth) degree classes. To deal with the problem of processing the uncertain rules, the next sections of this work will propose a generalized inference process and its implementation in a production system based on RETE networks.

2.1   Uncertain Predicates

Relations. First Order Logic is the logic of predicates, sentences stating properties of objects. The general form is $P(\mathbf{X})$, where P denotes a n-ary relation between the argument vector $\mathbf{X}$, with $|\mathbf{X}| = n$. According to the concrete values $\mathbf{x}$ of $\mathbf{X}$, the relation may hold or not in some degree: in general, a predicate is associated to a characteristic function

$$\mu_P : \mathbf{X} \to E$$

whose evaluation $\mu_P(\mathbf{x})$ states in which degree $\varepsilon(\mathbf{x}) \in E$ the given arguments satisfy the relation P. From now on, a predicate and its truth degree will be denoted using the compact syntax $P(\mathbf{X})_\varepsilon$. In the simplest case (n = 1) the single argument is typically an object and the predicate is used to state "x is P", i.e. that the relation In ($\in$) holds between x and the set defined by property P.

Truth degrees. Different definitions and interpretations exist for the set of truth values E. Usually $E = [0,1]$, as a single rational or real value is sufficient to represent any one type among probability, vagueness and confidence, provided that the degree is precise itself. When this condition is not satisfied, higher order truth degrees may be applied, including intervals ($E = [0,1]x[0,1]$, see [3]), fuzzy numbers ($E = 2^{[0,1]}$, see [8]) or, such as in the case of imprecise probabilities, high order fuzzy sets ($E = (2^{[0,1]})^m$, see [6]).

2.2   Generalized Operators

Operators are n-ary relations $Op(\mathbf{X})$ associated to characteristic functions evaluating a property of a set of other predicates: $\mathbf{X} = \left\{ P_j(\mathbf{X}_j)_{\varepsilon_j} \right\}_{j:1..n}$. For example, when P = "And", the property holds in a degree which is stronger the stronger are the degrees of all the argument predicates.

In general, the computed truth degree is function of both the truth degrees and the arguments of the evaluated predicates: $\varepsilon_{Op} = \mu(\{\mathbf{X}_j, \varepsilon_j\}_{j:1..n}) \in E$. This happens, for example, when a probabilistic conjunction $A \cap B$ is evaluated and A and B are not independent. More often, however, the operators are truth-functional: the evaluation then depends only on the truth degrees and not on the arguments: $\varepsilon_{Op} = \mu(\{\varepsilon_j\}_{j:1..n})$, so a general functional definition can be provided. Truth functionality is a desirable property for computational simplicity and efficiency, especially in the evaluation of nested operators forming complex expressions.

Operator families. In classical FOL, the set of binary operators is limited to conjunction, disjunction, implication and negation. It is known that any two of them are sufficient to define the others (e.g. negation and conjunction or negation and implication) and that, adding constants and derived operators, there can be no more than 2^4 different functions $0, 1 \times 0, 1 \rightarrow 0, 1$.

Their generalization in uncertain logic, however, has led to a much vaster set of operators: not only it depends on the type of uncertainty that is being modelled (probabilistic conjunction is different from vague conjunction) and the chosen truth degree set E, but even then there may be possible alternatives. However, since they are extensions of the crisp case, the mutual correlations still hold, so it is more correct to speak of operator families made of logically coherent definitions. In the fuzzy case, for example, it has been shown ([13]) that there are three different basic definitions (Godel, Gouguen and Lukasiewicz), which are the generators of a parametric class of operators families ([8]).

So, each abstract operator is associated to a class of evaluation functions, which, if truth-functional, are called "norms":

$$\neg : \text{Negation} \quad \Rightarrow \quad \text{c-norm}$$
$$* : \text{Conjunction} \Rightarrow \quad \text{t-norm}$$
$$+ : \text{Disjunction} \quad \Rightarrow \quad \text{u-norm}$$
$$\rightarrow: \text{Implication} \quad \Rightarrow \text{r-implication}$$

2.3 Formulas and Rules

For pure evaluation purposes, formulas can be formed according to the following abstract grammar:

```
Formula  ::= Predicate | Operator
Operator ::= OpName(OpArgs)
OpArgs   ::= Formula | Formula, OpArgs
```

Provided that each predicate is associated to an evaluation function and each operator is correctly defined, a pre-order visit of the tree allows a complete evaluation of the truth degree of a formula.

Rules themselves are special, usually general, formulas:

```
Rule       ::= Premise ImplicOp Conclusion
Premise    ::= Formula
Conclusion ::= Formula
```

Usually, formal systems require the Conclusion to be atomic, while this grammar allows a complex operator to appear in the right side of a rule, like in $Op_1(P_j) \rightarrow Op_2(C_k)$ (arguments are omitted for brevity). This is not a limitation as long as Op_2 is considered a predicate itself and no attempt is made to recover the truth value of the individual predicates C_k. Actually, in uncertain logic projection-like operations can be defined, some of which recall the way a belief measure over a set is divided among its elements, but this topic will not be explored further in this work. Instead, it is interesting to notice that a complex conclusion may be a sub-tree in a more complex formula $Op_3(Q_1, \ldots, Op_2(C_k), \ldots, Q_n)$, the evaluation of which might be speeded up if the high-level degree is directly available.

This also an important option for dynamic rules: the truth degree of a rule may be computed directly at run-time instead of being passed as a fact:

$$A \rightarrow (P \rightarrow C)$$

In this case, in fact, the consequence of the outermost implication is the truth degree of the second implication.

2.4 Inference Rules

Modus Ponens. Almost all rule-based systems rely on Modus Ponens, which uses a premise P and implication $P \rightarrow C$ to entail some conclusion C. Uncertain Modus Ponens uses an extended schema:

$$\frac{< P(\mathbf{x})_{\varepsilon_P}, P(\mathbf{X}) \rightarrow_{\varepsilon_\rightarrow} C(\mathbf{Y}) >}{C(\mathbf{y})_{\rho(\varepsilon_P, \varepsilon_\rightarrow)}}$$

When an available fact matches an existing premise, i.e. the arguments unify under some substitution θ, the consequent formula is stated with a truth degree which depends on both the degrees of the premise and the implication itself, which concrete definition depends on the operator family being used (e.g. in fuzzy logic a t-norm, usually min).

Merging Information. In classical logic, once the truth of a fact has been asserted there is no need to restate it even if there can be different premises supporting it (unless it is required to keep track of which premises led to a conclusion). This is different in uncertain logic, since different sources could return different truth values: for this reason, a further combination/intersection operator, denoted by $\cap$, is needed. Several concrete techniques exist, from S-implications to Dempster-Shafer's information merging approaches ([4]). In general, the truth degree of a formula C can be obtained in several ways:

- From prior knowledge expressed as a fact
- By direct evaluation, possibly recursively (for complex formulas)
- By specialization of a more general formula
- By Modus Ponens using premises and implications

The information merge raises the issue of the order in which rule fire, since the premise of one may be the consequence of several others as in the rule set $\{P_1 \rightarrow C, P_2 \rightarrow C, C \rightarrow D\}$. In this case, the third rule should fire only once after the other two have fired and their result has been merged. The problem has been debated in [14] and is crucial if a rule has some side effects which are not idempotent.

Induction. Induction is an aggregation process for the evaluation of a general formula $P(\mathbf{X})$ from a set of more specific examples $P(\mathbf{x} \preceq \mathbf{X})$, for example obtained by deduction. If the formula is an implication, it can be used to build rule systems capable of learning gradual rules ([9]).

2.5   Abstract Rule Bases

The existence of classes of operators and inference mechanisms leads to an interesting conclusion: a knowledge base exists at two levels of specification. The first is conceptual and abstract and is expressed by the qualitative, declarative syntax of the rules; the second is concrete and quantitative and depends on the actual definition and implementation of the operators. The main issue is coherence: the operators should belong to the same family or there is the risk of obtaining logically inconsistent results. Moreover, the operators should be compatible with the chosen type of truth degrees.

Fuzzy RuleML ([5]), exploiting the native tree structure of an XML document, supports complex formulas with nested operators and allows the customization of truth degrees and operators at a fine level of granularity, mainly thanks to the tag Degree and the attribute @kind. The next section will propose how to introduce the same flexibility in a RETE-based inference engine.

3   Introducing Uncertainty into RETE-OO Networks

So far, uncertain rule bases are nothing more than a set of formulas annotated with truth degrees to provide prior information and functions for the evaluation of operators. Their execution of can be performed in two canonical ways:

either by interpretation, using a general purpose rule engine, or by compilation, translating the rules into a structure capable of elaborating the information. The RETE algorithm uses this second approach and is widely used for its efficiency ([7]) and robustness in many expert system shells, such as Drools [11], Jess [2], Clips [1], and others.

RETE networks The RETE algorithm is well known and many references can be found, so here it will only recalled briefly. Given a rule in any of the equivalent forms "If P Then C", "When P Then C" or "P → C", the premise formula P expresses a set of constraints which must be matched by one or more objects before the rule can fire and enact the consequences C. When RETE is used, a set of rules is compiled into a network which nodes form a directed, acyclic graph. The nodes are complex entities capable of evaluating constraints and storing information: in order to activate the rules, the facts traverse the different nodes, possibly being combined with other facts. There are two main types of nodes:
    these constraints can be Select-like, as they filter single objects, or Join-like, when used to match different objects: the former are evaluated by the Alpha-Network, the latter by the Beta-Network.

- α-**Nodes**, which evaluate constraints on a single fact, similarly to a SELECT operation on a database table. α-Nodes are chained in sequence: the facts which satisy all the required constraints are stored in an α-Memory at the end of the chain.
- β-**Nodes**, which evaluate constraints on pairs of facts, similarly to a JOIN operation between two database tables. In general, a fact from an α-Memory is joined to a sequence of joined facts, called *tuple* and stored in a β-Memory, to yield a new tuple. As the tuple traverses the sequence of beta nodes, it grows adding new facts until it reaches a terminal node, activating a rule.

Objects vs Predicates. Given the greater diffusion of object-oriented languages such as Java, when compared to declarative logic-based ones, the concept of logic program has been adapted. An atom is no longer a predicate, but an object with an interface exposing some properties and their values. The original n-ary relation becomes a complex formula, which can typically expressed in predicate logic, composing relations on the property values. For example, consider the example expressed in the DRL language of Drools ([11]):

$$\text{p : Person(name == "Davide" , age} < 30 \text{ || height} > 180)$$

The formula accepts an object if and only if it models a person called Davide which is either young or tall, i.e. it is of class Person, and some of its fields satisfy the imposed constraints. In predicate logic, this could be expressed by any formula equivalent to the following, where $*$ denotes a conjunction, $+$ a disjunction and the symbol p denotes a variable reference:

$$\text{Person(p)} * (\text{Name(p,Davide)} * (\text{Younger(p,30)} + \text{Taller(p,180)}))$$

An equivalent formulation is:

Inst(p,Person) * (Eq(p.name,davide) * (Lt(p.age,30) + Gt(p.height,180)))

This version stresses the role of the evaluators hidden behind the definition of the predicates used in the previous version. In order to obtain an uncertain evaluation, it is sufficient to change their underlying implementation to return a truth degree instead of a simple Boolean. In the example, the subformula (Lt(p.age,30) + Gt(p.height,180)) evaluating the physical features of the person could be provided by a complex evaluator, possibly encapsulating an image analyzer fed with a picture of the candidate author. Notice that this evaluator would compute the truth degree of an *or* operator for a very specific subset of arguments. This requires the addition of the identity:

PersonEval(P,[tgtAge,tgtHgt]) $\equiv_1$ (Lt(P.age,tgtAge) + Gt(P.height,tgtHgt))

So, given the equivalence between predicate logic and object logic, the abstract inference schema outlined in section 2 will be applied to an object-oriented version of the RETE networks.

Alpha-Network

Constraint nodes. Generalizing the previous examples, in the ith rule, constraints over an object are usually specified using a conjunction of properties to be satisfied:

$$obj : Class(Eval_1(field_1,args[]_1), \ldots, Eval_n(field_n,args[]_n))$$

Each simple constraint $C_{i,j}$ has a signature σ_j identifying it univocally, at the point that identical constraints used in different rules are shared in the network. The signature can be obtained, for example, by combining the hash codes of the evaluator, fieldName, args tuple. If a constraint does not depend on other objects, it is mapped to a Select node: the node encapsulates the evaluator function $\varphi_j(obj.field, args[])$, which returns a truth degree ε_j^σ. Evaluators may be as simple as a crisp comparison operator or as complex as a neural network, as long as they satisfy the abstract definition. The evaluation of cross-object conditions, instead, are delayed until the Join Nodes in the Beta Network .

Combining priors. In order to introduce and combine prior information, each individual object has an associated Constraint Map<Signature,Degree>, called Priors(Σ), associating to each constraint σ_j the currently evaluated truth degree ε_j. The map is created when the object is inserted into the network for the first time and filled with the factual information π_j extracted from the logic program: Priors(σ_j) = ε_j = π_j. When the object passes through Select node j, the evaluator returns the degree ε_j^σ: this value must be merged with the available priors (if any) using an aggregation operator:

$$\varepsilon_j = \varepsilon_j \cap \varepsilon_j^\sigma$$

If the same object may be inserted several times, the contribution of the evaluator must be taken into account only once, so a flag $eval_j$ must be used to inhibit the activation of the evaluators. This can also be beneficial for the performances of the system if a custom evaluator is particularly demanding.

Object-level truth degree. The n conditions imposed on an object, including the class check, correspond to a sequence of m select-nodes $C_{i,1..m}(m \leq n)$, possibly shared with other rules, which implement some of the n operands of a customizable aggregation operator (the remaining $n - m$ constraints are join conditions). The sequence is translated into a path from the Entry node to the corresponding Alpha-Memory of length $m + 1$: the first m are Select nodes, while the last is an Operator node wrapping the specific combination strategy.

While individual constraints are node-dependent, the global $\varepsilon_{i,j}$ is path-dependent and can be computed only after n evaluators have been applied. During the traversal, a stack-like structure T is used: Select nodes push their result, while Operator nodes push a reference to the operator they implement. If an n-ary operator has a sequence of n truth degrees below itself, it pops itself and the n values, combines them and pushes the result.

Complex premises. If more than one operator is used in the formula, the Alpha path is a tree linearized by a pre-order visit. The only difference with the basic case is that more than one Operator node appears in the path.

Extended Alpha Network algorithm. The introduction of uncertainty does not alter the topology of the Alpha Network, except for the introduction of operator nodes, but changes the inner structure of the nodes, adding custom evaluators, and propagates additional information along with the objects. The traversal of the Alpha Network can be summarized according to the algorithm 3.

After the passage through the Alpha Network, an object has its Priors table updated with the evaluation of all the constraints in the paths it has traversed. The object is then placed in one or more Alpha Memories, each with a different, possibly partially computed truth degree.

Unlike the original architecture, where select nodes blocked objects for which the inner constraint evaluated to false, in the uncertain version the objects always reach the Alpha Memory. The presence of operators other than the conjunction, negation and exclusive or among others, allows the exploitation of false facts as well as true ones. An extended Alpha Memory, however, encapsulates a customizable filtering criterion that blocks useless objects from being passed into the Beta Network. The concrete definition depends on the adopted truth degree and may include:

- Block if truth degree too low (fuzzy truth values)
- Block if probability of truth too low (probabilistic truth values)
- Block if certainty factor too low (confidence truth values)
- Block if interval too wide (truth intervals)

Algorithm 1. Alpha Network Traversal

---

for all Fact $f = \{\sigma_j, \pi_j\}$ **do**
  obj.Priors$[\sigma_j] = \pi_j$
end for
Node = Rete.*entryNode*
while Node.next() != null **do**
  Node = Node.next()
  if Node.type == Node.SELECT **then**
    σ = Node.signature();
    if eval$_j$ **then**
      obj.Priors$[\sigma]$ = obj.Priors$[\sigma]$ $\cap$ Node.eval(obj)
    end if
    T.push(obj.Priors$[\sigma]$)
  else {Node.type == Node.OPERATOR}
    if there are Node.arity truth degrees on the top of T **then**
      operands = T.pushN(Node.arity)
      T.push(Node.eval(operands))
    else
      T.push(Node.op)
    end if
  end if
end while

- Block if entropy too high (truth distributions)
- Combinations of the above criteria (more complex representations)

If a truth degree has been computed partially because some join constraints are still to be evaluated, it should not be filtered. However, if the adopted concrete logic models the concept of Unknown, it may be substituted to the missing values for a heuristic computation to guide in the decision.

Beta-Network. The Beta Network is responsible of evaluating the relational matches among different objects: a join is the evaluation of a constraint involving properties of two objects. For example, the following rule premise is satisfied by a pair of objects modelling a paper and its author:

p : Person(age < 30 || height > 180), x : Paper(topic = "Drools" , author = p)

Each Person in the (left) memory has an associated truth degree; the papers, instead, have a partially evaluated degree of the form $T = [\varepsilon_2, *]$. Once a tentative pair has been formed, the missing constraint can be evaluated, and so the truth degree associated to each paper (given its candidate owner) can be computed and combined with the person's.

Join Sequences. Given a rule involving k objects, $k - 1$ join nodes are needed. Provided that the memories for each class contain $N_{i:1..k}$ objects, up to $\prod N_{i:1..k}$ tuples can be formed; each tuple contains k objects and their k truth values. In particular, join nodes must:

Algorithm 2. Beta Join Node Traversal

Require: new Tuple t in node.LeftMemory (new Object o in node.RightMemory)
  for all Object o in node.RightMemory (Tuple t in node.LeftMemory) **do**
    $\varepsilon_o \leftarrow$ node.constraint.evaluate(t,o);
    $t \leftarrow t.$append(o)
    if ! filter(t) **then**
      propagate(t)
    end if
  end for

When an object enters a Join node from the Right memory, the node contains the $n - m$ constraints still to be evaluated. The result of the evaluations, ordered according to a depth-first visit of the formula tree, are pushed into the partially full stack T from the bottom. After each insertion, the first operator with enough arguments is applied and the result is reinserted in T, activating the procedure recursively. In the example above, ε_1 is added to the stack of truth degrees of a Paper, yielding the evaluable $T = [\varepsilon_1, \varepsilon_2, *]$.

Complex Beta Network and Modus Ponens. In the standard RETE architecture, the chain of Join nodes and Beta Memories ends in a Terminal Node: in order to support uncertainty, the structure has to be extended. The output of the last Join Node of a rule premise is connected to an Operator Node, which combines the individual truth degrees of the objects in the argument tuple (possibly applying a filtering criterion, if enabled). In turn, the output of the Operator Node is connected to an Alpha Memory containing logical facts, i.e. formulas and their truth degrees. This memory allows the network to support complex premises, defined using nested operators: the join sequence required by an operator at level $l + 1$ can be attached directly to the sequence of an operator at level l.

The second, and more important, reason for this extension is the support for rules based on uncertain implications. The information on a rule Premise is beta-joined with the tuple denoting the rule itself, stored as a logical fact in a separate Alpha Memory. The truth degree of rules is usually given as a fact in the logic program, so it is possible that the Alpha Memory contains but a constant. However, if the rule is expressed using an implication and its degree has been evaluated dynamically from facts, the Alpha memory can be attached to the output of a beta operator node.

At this point, the pair [Premise, Rule] can be passed to an Operator implementing the adequate Modus Ponens strategy to obtain the truth degree of the consequence. Finally, the arguments of the Premise and the truth degree of the consequence can be passed to a Terminal Node which schedules the activation of the rule in the Agenda with the given parameters.

Logical Assertions. When a rule fires, its right-hand side is evaluated and the sequence of operations listed therein is executed. The consequence of a rule may include:

1. Statement of new facts
2. Justification of either new or existing facts
3. Non-logical operations, possibly with side-effects (not present in purely logical programs)

As far as the RETE network is concerned, only the first two possibilities are relevant. The truth degree obtained by Modus Ponens can be used in new facts about properties of objects or even about complex formulas.

Predicate-level assertions The assertion of a new object obj is the typical way of propagating information between rules in an object-based production system. The available truth degree may be used to provide information on one or more alpha-constraints on objects of obj's class, effectively building a Prior table for obj. The main issue is that obj (or an object equal to obj) may need to be asserted more than once: this is the case when different rules are used to assess different properties of the same object. In general, evaluators should be activated only the first time, while afterwards the new information should be merged with the existing one. To give greater flexibility, two operations are provided: State and Justify.

A Statement is the absolute assertion of a fact. It merges priors with evaluators, overriding all existing information (if any), and can not be modified except by a successive Statement.

A Justification is a weaker form of logical assertion: the first justification merges priors with evaluators, while the following ones just merge the information without triggering the evaluators. Moreover, a justification keeps track of which activations have contributed to the current result. Notice that the justification of a stated object is ignored.

The policy is qualitatively summarized in algorithm 3.

Formula-level assertions. Consider the constraint "o : Object(Eval$_j$(field$_j$, args$_j$))": a deduction might allow one to know that an object o satisfies the constraint globally in some degree ε without knowing the degree of satisfaction of the individual field constraints - or worse, without knowing the values o some fields of o. In this case, the pair o, ε has to be inserted directly in the Alpha Memory, bypassing the Alpha Network completely with the benefit of avoiding the issues described in the previous paragraph. Modus Ponens could also be used to infer the truth degree of a formula involving several objects, such as a conjunction or, more importantly, an implication. The Left Memories added to the Beta Network allow a uniform handling of such cases: if the asserted formula does not exist in the Alpha Memory, it is inserted therein; otherwise a Statement overrides the existing value while a Justify causes the old and the new value to be merged according to the current strategy $\cap$.

3.1   Case Study: Drools

Drools [11] is a highly configurable, java-based, open-source engine based on the RETE-OO algorithm. The proposed architecture is being implemented as

---

Algorithm 3. State/Justify Policy

---

if Object o has not been asserted before **then**
  if State **then**
    o.tag = STATE
    assert(o, priors, *eval* ← true, *merge* ← false)
  else {Justify}
    o.tag = JUSTIFY
    eval = true
    merge = false
    assert(o, priors, *eval* ← true, *merge* ← false)
  end if
else { Object o' exists s.t. o equals o'}
  if State **then**
    clear o'
    o.tag = STATE
    assert(o, priors, *eval* ← true, *merge* ← false)
  else {Justify}
    if o'.tag == JUSTIFY **then**
      assert(o, priors, *eval* ← false, *merge* ← true)
    else {o'.tag == STATE}
      Ignore justification
    end if
  end if
end if

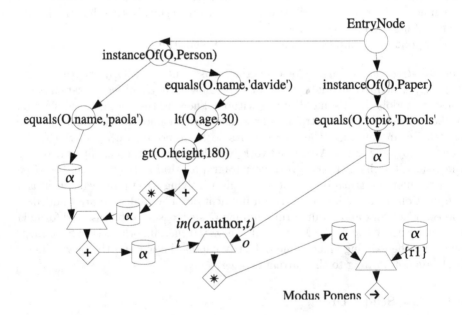

Fig. 1. Extended RETE network

an extension of the Drools core, with the goal of adding the support of different families of uncertain reasoning. It offers the concept of Session, in which a general RuleBase is used for inference over a particular set of objects. The rules are defined with abstract operators and truth values, which become concrete in different Sessions according to the setting of the properties Session.OPS (for operators) and Session.DEG (for truth degrees): they refer the appropriate Factories responsible for instantiating the actual evaluation behaviour. The interfaces *IOperatorFactory* and *IDegreeFactory* have been defined and different classes have been implemented to test the system: ConfidenceOpFactory, FuzzyOpFactory, RealDegFactory, IntervalDegFactory, DistributionDegFactory are provided as part of the bundle. The objects they create, embedded in the Rete network, compute the activation degree of the rules, which in turn is made available to the consequence part in the variable Conclusion.DEGREE. The comparison operators are being replaced by pluggable, customizable evaluators with user-defined behaviour. Even if uncertainty is optional and the system can still be used in the traditional way, the symbol $\sim$ has been adopted to enable the uncertain evaluation of a property. Further information can be specified with the use of attributes: at the moment, the attribute *degree* is being used to provide the initial truth degree of a rule, as in the example below.

rule "rule1"
degree ε_{r1}
when

      (p1 : Person(name $\sim=$ 'davide' , (height $\sim=$ 185 || weight $\sim=$ 75))
      Ops.OR p2 : Person(name $\sim=$ 'paola')
) Ops.AND Paper(topic $\sim=$ 'Drools', author in [p1,p2])
then

      print('Warning : The truer, the more this rule disables itself')
      insertUncertain('rule1',Ops.NOT(Conclusion.DEGREE))

4   Conclusions and Future Works

The extensions of fact-driven, forward logical inference necessary in presence of uncertainty have been integrated in the Rete network, a standard implementation of inference engines. The proposed architecture is being integrated into the core of the Drools rule engine, to give a transparent support of uncertain rules without altering the way standard rules perform: for the system to be usable, some extensions to the rule language are necessary to express the additional concepts; in the same context, compatibility with RuleML will be considered. The study will also be extended to the case of non-monotonic reasoning, in order to support the retraction and the modification of facts, and to the development of specific conflict resolution strategies which exploit the actual activation degrees of the rules.

References

1. Clips, a tool for building expert systems, http://clipsrules.sourceforge.net/
2. Jess, the rule engine for java platform, http://www.jessrules.com/
3. Atanassov, K.T., Koshelev, M., Kreinovich, V., Rachamreddy, B., Yasemis, H.: Fundamental justification of intuitionistic fuzzy logic and of interval-valued fuzzy methods
4. Cuzzolin, F.: Geometry of dempster's rule of combination. IEEE Transactions on Systems, Man, and Cybernetics, Part B 34(2), 961–977 (2004)
5. Damásio, C.V., Pan, J.Z., Stoilos, G., Straccia, U.: An approach to representing uncertainty rules in ruleML. In: Eiter, E.T., Franconi, R., Hodgson, S. (eds.) RuleML, pp. 97–106. IEEE Computer Society, Los Alamitos (2006)
6. Denoeux,: Reasoning with imprecise belief structures. IJAR: International Journal of Approximate Reasoning 20 (1999)
7. Doorenbos, R.B.: Production matching for large learning systems. Technical Report CS-95-113, Carnegie Mellon University, School of Computer Science
8. Dubois, D., Prade, H.: Fuzzy Sets and Systems: Theory and Applications. Academic Press, London (1980)
9. Dubois, D., Hüllermeier, E., Prade, H.: A systematic approach to the assessment of fuzzy association rules. Data Min. Knowl. Discov 13(2), 167–192 (2006)
10. Boley, H., et al.: Fol ruleml: The first-order logic web language, http://www.ruleml.org/fol/
11. Proctor, M., et al.: Drools, http://www.jboss.org/drools/
12. Forgy, C.: Rete: A fast algorithm for the many patterns/many objects match problem. Artif. Intell 19(1), 17–37 (1982)
13. Hájek, P.: Metamathematics of Fuzzy Logic. Trends in Logic: Studia Logica Library, vol. 4. Kluwer Academic Publishers, Dordrecht (1998)
14. Hall, L.O.: Rule chaining in fuzzy expert systems. IEEE-FS 9, 822–828 (2001)
15. Pan, J.Z., Stamou, G.B., Tzouvaras, V., Horrocks, I.: f-SWRL: A fuzzy extension of SWRL. In: Duch, W., Kacprzyk, J., Oja, E., Zadrożny, S. (eds.) ICANN 2005. LNCS, vol. 3697, pp. 829–834. Springer, Heidelberg (2005)

Programming with Fuzzy Logic Rules
by Using the FLOPER Tool*

Pedro J. Morcillo and Gines Moreno

University of Castilla-La Mancha
Department of Computing Systems
02071, Albacete (Spain)
{pmorcillo,gmoreno}@dsi.uclm.es

Abstract. The "Fuzzy LOgic Programming Environment for Research", FLOPER in brief, that we have implemented in our research group, is intended to help the development of rule-based applications supporting fuzzy logic and approximated reasoning. The system is able to directly translate a powerful kind of fuzzy logic programs (belonging to the so-called *multi-adjoint logic approach*) into Prolog code which can be directly executed inside any standard Prolog interpreter in a completely transparent way for the final user. The system also generates a low-level representation of the fuzzy code offering debugging (tracing) capabilities with close connections to other program manipulation tasks (optimization, specialization, etc). Our approach focuses on practical and technical aspects on rule-based reasoning with uncertain and fuzzy information.

Keywords: Fuzzy Logic Programming, Rule-Based Environments.

1 Introduction

Logic Programming [9] has been widely used for problem solving and knowledge representation in the past. Nevertheless, traditional LP languages do not incorporate techniques or constructs to treat explicitly with uncertainty and approximated reasoning. To overcome this situation, during the last years several fuzzy logic programming systems have been developed where the classical inference mechanism of SLD–Resolution has been replaced with a fuzzy variant able to handle partial truth and to reason with uncertainty [2,3,5,11].

Informally speaking, in the so-called multi-adjoint logic framework of [11], a program can be seen as a set of rules each one annotated by a truth degree, and a goal is a query to the system, i.e., a set of atoms linked with connectives called *aggregators*. A *state* is a pair $\langle \mathcal{Q}, \sigma \rangle$ where $\mathcal{Q}$ is a goal and σ a substitution (initially, the identity substitution). States are evaluated in two separate computational phases. Firstly, *admissible steps* (a generalization of the classical *modus ponens* inference rule) are systematically applied by a backward reasoning

---

* This work has been partially supported by the EU (FEDER), and the Spanish Science and Education Ministry (MEC) under grants TIN 2004-07943-C04-03 and TIN 2007-65749.

N. Bassiliades, G. Governatori, and A. Paschke (Eds.): RuleML 2008, LNCS 5321, pp. 119–126, 2008.

procedure in a similar way to classical resolution steps in pure logic programming, thus returning a computed substitution together with an expression where all atoms have been exploited. This last expression is then interpreted under a given lattice, hence returning a pair ⟨*truth degree*; *substitution*⟩ which is the fuzzy counterpart of the classical notion of computed answer traditionally used in LP.

The main goal of this paper is the detailed description of the FLOPER system (see a preliminary introduction in [1]) which is freely available via the web page: http://www.dsi.uclm.es/investigacion/dect/FLOPERpage.htm. Nowadays, the tool provides facilities for executing as well as for debugging (by generating declarative traces) such kind of fuzzy programs, thus fulfilling the gap we have detected in the area. Our implementation methods are based on two different, almost antagonistic ways (regarding simplicity and precision features), for generating pure Prolog code, with some correspondences with other previous attempts described in the specialized literature, specially the one detailed in [3].

The outline of this work is as follows. In Section 2 we detail the main features of multi-adjoint logic programming, both syntax and procedural semantics. Section 3 explain how to execute and debug such programs inside our FLOPER tool, which nowadays is being equipped with new options for performing other advanced program manipulation tasks (transformation, specialization, optimization). The benefits of our approach are highlighted by contrasting them with some related works in Section 4. Finally, in Section 5 we present our conclusions and propose some lines of future work.

2   Multi-adjoint Logic Programs

In what follows, we present a short summary of the main features of our language (we refer the reader to [11] for a complete formulation). We work with a first order language, $\mathcal{L}$, containing variables, function symbols, predicate symbols, constants, quantifiers ($\forall$ and $\exists$), and several (arbitrary) connectives to increase language expressiveness. In our fuzzy setting, we use implication connectives ($\leftarrow_1, \leftarrow_2, \ldots, \leftarrow_m$) and also other connectives which are grouped under the name of "aggregators" or "aggregation operators". They are used to combine/propagate truth values through the rules. The general definition of aggregation operators subsumes conjunctive operators (denoted by $\&_1, \&_2, \ldots, \&_k$), disjunctive operators ($\vee_1, \vee_2, \ldots, \vee_l$), and average and hybrid operators (usually denoted by $@_1, @_2, \ldots, @_n$). Although the connectives $\&_i, \vee_i$ and $@_i$ are binary operators, we usually generalize them as functions with an arbitrary number of arguments. By definition, the truth function for an n-ary aggregation operator $[\![@]\!] : L^n \to L$ is required to be monotone and fulfills $[\![@]\!](\top, \ldots, \top) = \top$, $[\![@]\!](\bot, \ldots, \bot) = \bot$. Additionally, our language $\mathcal{L}$ contains the values of a multi-adjoint lattice, $\langle L, \preceq, \leftarrow_1, \&_1, \ldots, \leftarrow_n, \&_n \rangle$, equipped with a collection of adjoint pairs $\langle \leftarrow_i, \&_i \rangle$, where each $\&_i$ is a conjunctor intended to the evaluation of *modus ponens*. In general, the set of truth values L may be the carrier of any complete bounded lattice but, for simplicity, in this paper we shall select L as the set of real numbers in the interval $[0, 1]$.

A *rule* is a formula $A \leftarrow_i \mathcal{B}$, where A is an atomic formula (usually called the *head*) and $\mathcal{B}$ (which is called the *body*) is a formula built from atomic formulas $B_1, \ldots, B_n$ ($n \geq 0$), truth values of L and conjunctions, disjunctions and aggregations. Rules with an empty body are called *facts*. A *goal* is a body submitted as a query to the system. Variables in a rule are assumed to be governed by universal quantifiers. Roughly speaking, a multi-adjoint logic program is a set of pairs $\langle \mathcal{R}; v \rangle$, where $\mathcal{R}$ is a rule and v is a *truth degree* (a value of L) expressing the confidence which the user of the system has in the truth of the rule $\mathcal{R}$. Often, we will write "$\mathcal{R}$ with v" instead of $\langle \mathcal{R}; v \rangle$.

In order to describe the procedural semantics of the multi–adjoint logic language, in the following we denote by $\mathcal{C}[A]$ a formula where A is a sub-expression (usually an atom) which occurs in the –possibly empty– context $\mathcal{C}[]$ whereas $\mathcal{C}[A/A']$ means the replacement of A by A' in context $\mathcal{C}[]$. Moreover, $Var(s)$ denotes the set of distinct variables occurring in the syntactic object s, $\theta[Var(s)]$ refers to the substitution obtained from θ by restricting its domain to $Var(s)$ and $mgu(E)$ denotes the *most general unifier* of an equation set E. In the following definition, we always consider that A is the selected atom in goal $\mathcal{Q}$.

Definition 1 (Admissible Steps). *Let $\mathcal{Q}$ be a goal and let σ be a substitution. The pair $\langle \mathcal{Q}; \sigma \rangle$ is a state. Given a program $\mathcal{P}$, an admissible computation is formalized as a state transition system, whose transition relation $\rightarrow_{AS}$ is the smallest relation satisfying the following admissible rules (also called $\rightarrow_{AS1}$, $\rightarrow_{AS2}$ and $\rightarrow_{AS3}$):*

1) $\langle \mathcal{Q}[A]; \sigma \rangle \rightarrow_{AS} \langle (\mathcal{Q}[A/v\&_i\mathcal{B}])\theta; \sigma\theta \rangle$ *if* $\theta = mgu(\{A' = A\})$, $\langle A' \leftarrow_i \mathcal{B}; v \rangle$ *in* $\mathcal{P}$ *and $\mathcal{B}$ is not empty.*
2) $\langle \mathcal{Q}[A]; \sigma \rangle \rightarrow_{AS} \langle (\mathcal{Q}[A/v])\theta; \sigma\theta \rangle$ *if* $\theta = mgu(\{A' = A\})$, *and* $\langle A' \leftarrow_i; v \rangle$ *in* $\mathcal{P}$.
3) $\langle \mathcal{Q}[A]; \sigma \rangle \rightarrow_{AS} \langle (\mathcal{Q}[A/\bot]); \sigma \rangle$ *if there is no rule in $\mathcal{P}$ whose head unifies with A (this case copes with possible unsuccessful admissible derivations).*

Definition 2. *Let $\mathcal{P}$ be a program with an associated multi-adjoint lattice $\langle L, \preceq \rangle$ and let $\mathcal{Q}$ be a goal. An admissible derivation is a sequence $\langle \mathcal{Q}; id \rangle \rightarrow^*_{AS} \langle \mathcal{Q}'; \theta \rangle$. When $\mathcal{Q}'$ is a formula not containing atoms and $r \in L$ is the result of interpreting $\mathcal{Q}'$ in $\langle L, \preceq \rangle$, the pairs $\langle \mathcal{Q}'; \sigma \rangle$ and $\langle r; \sigma \rangle$, where $\sigma = \theta[Var(\mathcal{Q})]$, are called admissible computed answer (a.c.a.) and fuzzy computed answer (f.c.a.), respectively.*

In order to illustrate our definitions, consider now the following program $\mathcal{P}$ and lattice $([0,1], \leq)$, where $\leq$ is the usual order on real numbers.

$\mathcal{R}_1 : p(X) \leftarrow_P q(X, Y) \&_G r(Y)$ *with* 0.8 $\mathcal{R}_2 : q(a, Y) \leftarrow_P s(Y)$ *with* 0.7
$\mathcal{R}_3 : q(b, Y) \leftarrow_L r(Y)$ *with* 0.8 $\mathcal{R}_4 : r(Y) \leftarrow$ *with* 0.7
$\mathcal{R}_5 : s(b) \leftarrow$ *with* 0.9

The labels P, G and L mean for *Product logic*, *Gödel intuitionistic logic* and *Lukasiewicz logic*, respectively. That is, $[\![\&_P]\!](x,y) = x \cdot y$, $[\![\&_G]\!](x,y) = min(x,y)$, and $[\![\&_L]\!](x,y) = max(0, x+y-1)$. In the following admissible derivation for the program $\mathcal{P}$ and the goal $\leftarrow p(X)\&_G r(a)$, we underline the selected expression in each admissible step:

$$\langle \underline{p(X)}\&_G r(a); id\rangle \rightarrow_{AS1}{}^{R_1} \langle (0.8\&_P(\underline{q(X_1,Y_1)}\&_G r(Y_1)))\&_G r(a); \sigma_1\rangle$$
$$\rightarrow_{AS1}{}^{R_2} \langle (0.8\&_P((\overline{0.7\&_P s(Y_2)}))\&_G r(Y_2)))\&_G r(a); \sigma_2\rangle$$
$$\rightarrow_{AS2}{}^{R_5} \langle (0.8\&_P((0.7\&_P 0.9)\&_G \underline{r(b)}))\&_G r(a); \sigma_3\rangle$$
$$\rightarrow_{AS2}{}^{R_4} \langle (0.8\&_P((0.7\&_P 0.9)\&_G \overline{0.7}))\&_G \underline{r(a)}; \sigma_4\rangle$$
$$\rightarrow_{AS2}{}^{R_4} \langle (0.8\&_P((0.7\&_P 0.9)\&_G 0.7))\&_G 0.7; \sigma_5\rangle,$$

where $\sigma_1 = \{X/X_1\}$, $\sigma_2 = \{X/a, X_1/a, Y_1/Y_2\}$, $\sigma_3 = \{X/a, X_1/a, Y_1/b, Y_2/b\}$, $\sigma_4 = \{X/a, X_1/a, Y_1/b, Y_2/b, Y_3/b\}$, and $\sigma_5 = \{X/a, X_1/a, Y_1/b, Y_2/b, Y_3/b, Y_4/a\}$. So, since $\sigma_5[Var(\mathcal{Q})] = \{X/a\}$, the a.c.a. associated to this admissible derivation is: $\langle (0.8\&_P((0.7\&_P 0.9)\&_G 0.7))\&_G 0.7; \{X/a\}\rangle$. Now, after evaluating the first arithmetic expression (where all atoms have been solved), we obtain the final fuzzy computed answer (f.c.a.) $\langle 0.504; \{X/a\}\rangle$.

3   The FLOPER System

As detailed in [1], our parser has been implemented by using the classical DCG's (*Definite Clause Grammars*) resource of the Prolog language, since it is a convenient notation for expressing grammar rules. Once the application is loaded inside a Prolog interpreter (in our case, Sicstus Prolog v.3.12.5), it shows a menu which includes options for loading, parsing, listing and saving fuzzy programs, as well as for executing fuzzy goals. All these actions are based in the translation of the fuzzy code into standard Prolog code. The key point is to extend each atom with an extra argument, called *truth variable* of the form $\_TV_i$, which is intended to contain the truth degree obtained after the subsequent evaluation of the atom. For instance, the first clause in our target program is translated into: "p(X,_TV0) : −q(X,Y,_TV1),r(Y,_TV2),and_godel(_TV1,_TV2,_TV3), and_prod(0.8,_TV3,_TV0). ", where the definition of the "aggregator predicates" are: "and_prod(X,Y,Z) : −Z is X ∗ Y." and "and_godel(X,Y,Z) : −(X =< Y,Z = X;X > Y,Z = Y).". The last clause in the program, becomes the pure Prolog fact "s(b,0.9)." while a fuzzy goal like "p(X) &godel   r(a)", is translated into the pure Prolog goal: "p(X,_TV1),r(a,_TV2), and_godel(_TV1,_TV2,Truth_degree)" (note that the last truth degree variable is not anonymous now) for which the Prolog interpreter returns the two desired fuzzy computed answers [Truth_degree=0.504,X=a] and [Truth_degree=0.4,X=b].

The previous set of options suffices for running fuzzy programs: all internal computations (including compiling and executing) are pure Prolog derivations whereas inputs (fuzzy programs and goals) and outputs (fuzzy computed answers) have always a fuzzy taste, which produces the illusion on the final user of being working with a purely fuzzy logic programming tool.

However, when trying to go beyond program execution, our method becomes insufficient. In particular, observe that we can only simulate complete fuzzy derivations (by performing the corresponding Prolog derivations based on

SLD-resolution) but we can not generate partial derivations or even apply a single admissible step on a given fuzzy expression. This kind of low-level manipulations are mandatory when trying to incorporate to the tool some program transformation techniques such as those based on fold/unfold (i.e., contraction and expansion of sub-expressions of a program using the definitions of this program or of a preceding one, thus generating more efficient code) or partial evaluation we have described in [4,6,7]. For instance, our fuzzy unfolding transformation is defined as the replacement of a program rule $\mathcal{R} : (A \leftarrow_i \mathcal{B}$ with $v)$ by the set of rules $\{A\sigma \leftarrow_i \mathcal{B}'$ with $v \mid \langle \mathcal{B}; id \rangle \rightarrow_{AS} \langle \mathcal{B}'; \sigma \rangle\}$, which obviously requires the implementation of mechanisms for generating derivations of a single step, rearranging the body of a program rule, applying substitutions to its head, etc.

Apart from the compilation method commented before and reported in [1], we have conceived a new low-level representation for the fuzzy code which nowadays already offers the possibility of performing debugging actions such as tracing a FLOPER work session. For instance, after parsing the first rule of our program, we obtain the following expression (which is "asserted" into the database of the interpreter as a Prolog fact, but it is never executed directly, in contrast with the previous Prolog-based representation of the fuzzy code):

```
rule(number(1), head(atom(pred(p,1),var('X')])), impl('prod'),
     body(and('godel',2,[atom(pred(q,2),[var('X'),var('Y')]),
                    atom(pred(r,1),[var('Y')])]])),
     td(0.8)).
```

Two more examples: substitutions are modeled by lists of terms of the form link(V, T) where V and T contains the code associated to an original variable and its corresponding (linked) fuzzy term, respectively, whereas an state is represented by a term with functor state/2. We have implemented predicates for manipulating such kind of code at a very low level in order to unify expressions, compose substitutions, apply admissible/interpretive steps, etc.

FLOPER is equipped with two options, called "tree" and "depth", for tracing execution trees and fixing the maximum length allowed for their branches (initially 3), respectively. As we are going to illustrate with two simple examples, the new options will be crucial when the "run" option fails: remember that this last option is based on the generation of pure logic SLD-derivations which might fall in loop or directly fail in some cases, in contrast with the traces (based on finite, non-failed, admissible derivations) that the "tree" option displays. Firstly, consider the following program $\mathcal{P}_1$:

```
p(X) <prod q(X) with 0.9.          p(X) <godel r(X) with 0.8.
q(X) <luka q(X) with 0.7.          r(a) with 0.6.
```

For goal p(a), FLOPER displays the first tree (trace 1) showed in Figure 1. Observe that each node contains an state (composed by the corresponding goal and substitution) preceded by the number of the program rule used by the admissible step leading to it (root nodes and nodes obtained via $\rightarrow_{AS3}$ are always labeled with the virtual, non existing rule R0). Nodes belonging to the same branch appear in different lines appropriately indented to help the readability

```
% TRACE 1: Execution tree with depth 4 for goal p(a) w.r.t.
             the multi-adjoint logic program P₁.
R0 < p(a), {} >
  R1 < &prod(0.9,q(a)), {X1/a} >
      R3 < &prod(0.9,&luka(0.7,q(a))), {X1/a,X7/a} >
          R3 < &prod(0.9,&luka(0.7,&luka(0.7,q(a)))), {X1/a,.. } >
              R3 < &prod(0.9,&luka(0.7,&luka(0.7,&luka(0.7,q(a))))),. >
  R2 < &godel(0.8,r(a)), {X2/a} >
      R4 < &godel(0.8,0.6), {X2/a} >


% TRACE 2: Execution tree with depth 2 for goal p(X) w.r.t.
             the multi-adjoint logic program P₂.
    R0 < p(X), {} >
        R1 < &prod(0.9,@aver(1,p(b))), {X/a} >
            R0 < &prod(0.9,@aver(1,0)), {X/a} >
```

Fig. 1. Traces and execution trees generated by FLOPER

of the figure. In our case, the tree contains only two different branches. It is easy to see that the first one, corresponding to the first five lines of the figure, represents an infinite branch, whereas the second one, identified by lines 1, 6 and 7, indicates that the goal has just one solution with truth degree 0.6 (which is the result of evaluating the arithmetic expression $\&godel(0.8, 0.6)$) even when the run option would fall in loop.

Our second example is not involved with infinite branches, but it copes with other kind of (pure Prolog) unsuccessful behaviour. Consider now a fuzzy program, say P_2, containing the single rule "p(a) $<$ prod @aver(1,p(b)) with 0.9" (where the average aggregator @aver has the obvious meaning) which, once parsed by FLOPER, is translated into the following pure Prolog clause: $p(a, TV0) : -p(b, \_TV1), agr\_aver(1, \_TV1, \_TV2), and\_prod(0.9, \_TV2, TV0)$. It is easy to see that, in order to execute goal "p(X)" by means of the "run" option, the Prolog interpreter will fail when trying to solve the first atom, "p(b, _TV1)", appearing in the body of this Prolog clause (no SLD-resolution step is applicable here, since this atom does not unify with the head $p(a, TV0)$ of the unique clause in the final Prolog program). However, in the fuzzy setting we know that the proposed goal has a solution, as revealed by the (single) successful branch appearing in the second trace of Figure 1. By applying and admissible step of kind 3 (see $\rightarrow_{AS3}$ in Definition 1), on the second node of the tree, we generate the final state showed in the third line of the figure (the system simply replaces the non solvable selected atom "p(b)" by the lowest truth degree 0). Note that this last state, once evaluated the associated arithmetic expression, returns the fuzzy computed answer X = a with truth degree 0.45.

As we have seen, the generation of traces based on execution trees, contribute to increase the power of FLOPER by providing debugging capabilities which allow us to discover solutions for queries even when a pure Prolog compilation-execution process fails. For the future and also supported on the generation of execution trees, we plan to introduce new options into the FLOPER menu implementing all the transformation techniques we are proposed in the past [4,6,7]: the key point will be the correct manipulation of the leaves of this kind of partially evaluated trees, in order to produce unfolded rules, reductants, etc.

4   Related Work

The multi-adjoint logic approach and the fuzzy logic language described in [3] are very close between themselves, with a similar syntax based on "weighted" rules (some precedents can be found in [2,5]) and levels of flexibility and expressiveness somehow comparable. However, whereas in the fuzzy language presented in [3] truth degrees are based on Borel Algebras (i.e., union of intervals of real numbers), in the so called *multi-adjoint logic programming* approach of [10,11] truth degrees are elements of any given lattice. Other important difference between both languages emerges at an operational level, since the underlying procedural principle of the language of [3] introduces several problems when considering most of the transformation techniques we are developing in our group. As we detail in [4,6], the adequacy of this language for being used as the basis of fuzzy folding/unfolding rules (program optimization) is rather limited: the real problem does not appear only at the syntactic level, but what is worse, the major inconvenience is the need for redefining the core of its procedural mechanism to cope with constraints possibly mixed with atoms.

Focusing now in the multi-adjoint logic approach, it is unavoidable to mention the implementation issues documented in [10]. Like our proposal, we all deal with the same target ([0;1]-valued) multi-adjoint logic language[1], and also our developments are based in pure Prolog code (even when they are supported on a neural net architecture). However, whereas they are restricted to the propositional case, we have lifted our results to the more general first-order case. Moreover, the procedural semantics implemented in [10] has been conceived as a bottom-up procedure where the repeated iteration of an appropriately defined consequence operator reproduces the model of a program, thus obtaining the computed truth-values of all propositional symbols involved in that program (in a parallel way). In a complementary sense, the executing and debugging "query answering" procedures implemented in FLOPER, are goal-oriented and have a top-down behaviour.

5   Conclusions and Future Work

In this paper we were concerned with implementation techniques for fuzzy logic programming and more exactly, for the multi-adjoint logic approach, which

---

[1] We all focus in a simple lattice whose carrier set is the real interval $[0, 1]$ and the connectives are collected from classical fuzzy logics (as the product, Łukasiewicz and Gödel intuitionistic logic). An high-priority task for future developments will be to let our system accept fuzzy programs as well as multi-adjoint lattices in a parametric way, which implies the design of appropriate protocols, interfaces, etc.

enjoys high levels of expressivity and a clear operational mechanism. Apart from [10] and our prototype tool FLOPER (see a preliminary description in [1] and visit http://www.dsi.uclm.es/investigacion/dect/FLOPERpage.htm), there are not abundant tools available in practice, which justifies our proposal.

We have firstly proposed a technique for running such kinds of programs based on a "transparent compilation process" to standard Prolog code. Secondly, we have next proposed an alternative compilation way which produces a low-level representation of the fuzzy code allowing the possibility of debugging (by generating declarative traces) the execution of a given program and goal.

This last development also opens the door to implement new and powerful program manipulation techniques (fold/unfold-based program optimization, program specialization by partial evaluation, etc.), such as the ones we describe in [4,6,7]. These actions, together with the extension of the cost analysis proposed in [8], are some high-priority tasks in our research group for the near future.

References

1. Abietar, J.M., Morcillo, P.J., Moreno, G.: Designing a software tool for fuzzy logic programming. In: Simos, T.E., Maroulis, G. (eds.) Proc. of ICCMSE 2007. Computation in Modern Science and Engineering, vol. 2, pp. 1117–1120. American Institute of Physics (distributed by Springer) (2007)
2. Baldwin, J.F., Martin, T.P., Pilsworth, B.W.: Fril- Fuzzy and Evidential Reasoning in Artificial Intelligence. John Wiley & Sons, Inc., Chichester (1995)
3. Guadarrama, S., Muñoz, S., Vaucheret, C.: Fuzzy Prolog: A new approach using soft constraints propagation. Fuzzy Sets and Systems 144(1), 127–150 (2004)
4. Guerrero, J.A., Moreno, G.: Optimizing Fuzzy Logic Programs by Unfolding, Aggregation and Folding. In: Visser, J., Winter, V. (eds.) Proc. of the 8th. International Workshop on Rule-Based Programming, RULE 2007. Electronic Notes in Theoretical Computer Science, 15 pages. Elsevier, Amsterdam (to appear, 2007)
5. Ishizuka, M., Kanai, N.: Prolog-ELF Incorporating Fuzzy Logic. In: Joshi, A.K. (ed.) Proceedings of the 9th International Joint Conference on Artificial Intelligence, IJCAI 1985, pp. 701–703. Morgan Kaufmann, San Francisco (1985)
6. Julián, P., Moreno, G., Penabad, J.: On Fuzzy Unfolding. A Multi-adjoint Approach. Fuzzy Sets and Systems 154, 16–33 (2005)
7. Julián, P., Moreno, G., Penabad, J.: Efficient reductants calculi using partial evaluation techniques with thresholding. Electronic Notes in Theoretical Computer Science 188, 77–90 (2007)
8. Julián, P., Moreno, G., Penabad, J.: Measuring the interpretive cost in fuzzy logic computations. In: Masulli, F., Mitra, S., Pasi, G. (eds.) WILF 2007. LNCS (LNAI), vol. 4578, pp. 28–36. Springer, Heidelberg (2007)
9. Lloyd, J.W.: Foundations of Logic Programming. Springer, Berlin (1987)
10. Medina, J., Mérida-Casermeiro, E., Ojeda-Aciego, M.: A neural implementation of multi-adjoint logic programs via sf-homogeneous programs. Mathware & Soft Computing XII, 199–216 (2005)
11. Medina, J., Ojeda-Aciego, M., Vojtáš, P.: Similarity-based Unification: a multi-adjoint approach. Fuzzy Sets and Systems 146, 43–62 (2004)

Ruling Networks with RDL:
A Domain-Specific Language to Task Wireless Sensor Networks

Kirsten Terfloth and Jochen Schiller

Institute of Mathematics and Computer Science
Takustr. 9, 14195 Berlin, Germany
{terfloth,schiller}@inf.fu-berlin.de
http://cst.mi.fu-berlin.de/

Abstract. Events are a fundamental concept in computer science with decades of research contributing to enable precise specification and efficient processing. New as well as evolving application domains nevertheless call for adaptation of successful concepts to meet intrinsic challenges provided by the target environment. A representative of such a new area for application are wireless sensor networks, pushing the need for event handling onto the bare metal of embedded devices. In this paper, we motivate the deployment of reactive rules in wireless sensor networks and describe our rule-based language RDL. Since our goal is to provide a high level of abstraction for node-level tasking, we will especially focus on recent additions to the language that support modularity to achieve a better encapsulation of concerns.

Keywords: domain-specific language, reactive rules, RDL, FACTS.

1 Introduction

One major application area of the rule-based programming paradigm are event-centric systems. Here, the flow of program execution seldom follows a linear, predefined path but is rather asynchronously determined at runtime by incoming events to be instantly processed by the system. Mapping exactly this system behavior is facilitated when using a language that has been designed to specify conditions denoting the emergence of a - possibly composite - event and actions performed in case of event recognition as opposed to using a general-purpose language. Therefore, rule-based languages with their inherent condition-action structure are a natural solution to orchestrate event-centric processing.

A rather young field of application for event-processing are wireless sensor networks. The idea is to use these networks as a decentralized tool of interconnected embedded devices that autonomously observe their environment and either report sampled data on a periodical basis or in an event-based manner. A wireless transceiver, a processing entity and a set of application-dependent sensors and actors on the sensor nodes enable each device to act as smart data source, a filter, an aggregator or a router to name but a few possible roles.

N. Bassiliades, G. Governatori, and A. Paschke (Eds.): RuleML 2008, LNCS 5321, pp. 127–134, 2008.
© Springer-Verlag Berlin Heidelberg 2008

Programming a wireless sensor network is however not a trivial task: Aside from having to deal with asynchronous processing, an application developer faces embedded devices as well as nodal distribution. To make this domain accessible beyond sensor network experts, thus to overcome the predominantly system-oriented point of view developers have to adopt, abstractions from low-level system concerns are mandatory.

In this paper, we address this need for abstraction by providing an overview of our programming language RDL (ruleset definition language) that allows to specify rules for embedded event processing. Their inherent modularity, declarative nature and high level of abstraction to especially capture event semantics make rules a perfect choice for utilization in wireless sensor networks. This paper extends our work on RDL presented in [1] in several significant ways: Earlier work has primarily provided a first rationale on the language. Here, we introduce a language that has been proven useful in a number of application scenarios, e.g the implementation of distributed event detection on a construction fence, and thus matured during this process. Specifically, we describe a set of additions to RDL that enable the definition of modules to ease code reuse. We argue that the usage of such *rulesets* as a means for encapsulation of semantically interacting rules facilitates application development and improves overall utilization.

The remainder of this paper is organized as follows: In Section 2 we give an overview of the architecture of our middleware framework FACTS which enables devices to interpret rules. Section 3.1 reviews language syntax and semantics of RDL. Extensions to the core language for supporting modularity and scoping are presented an discussed in Section 3.2, before pointing out work related to ours and concluding the paper.

2   The Big Picture

FACTS is a lightweight middleware framework developed to run on embedded networked sensors. It enables interpretation of rules expressed in RDL on sensor nodes, thus combines the advantages of a high-level programming abstraction and a sandboxed execution environment. Manual memory management and concurrency are hidden and pushed into the FACTS framework. A programmer can depend on a simple, data-centric paradigm to formalize events of interest, conditions for rule execution and corresponding actions taken upon event recognition. Rules are supplied to the rule engine as compiled bytecode, which allows for heavy optimizations concerning bytecode size. Both, the language RDL and the FACTS framework have been designed specifically to deal with the harsh memory constraints of embedded devices: Many design decisions, such as the absence of local variables in RDL, the chosen pattern-based strategy for rule evaluation and syntax and semantics of available actions in a rule contribute to this challenge. An implementation of FACTS is available for ScatterWeb sensor nodes which feature solely 5 KB of RAM. The software architecture, including a rule engine to process deployed rulesets and a fact repository serving as the central working memory of the node, is depicted in Figure 1.

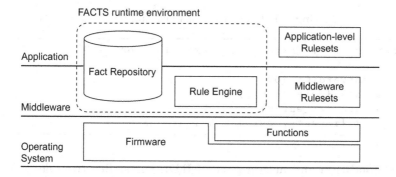

Fig. 1. Components of the FACTS middleware framework

3   RDL Language Details

Syntax and semantics for specifying cause and effect in RDL are tailored to meet sensor network requirements rather than to offer a multi-purpose rule-based language. To provide a first impression, we will start off with an informal discussion of a small example for transparent data forwarding, see Listing 1.1.

RDL relies on a simple, yet versatile model for data specification, so called facts, which are named tuples of typed values. Line 1 e.g. specifies a fact *rt_entry*, representing an entry in a routing table, that features three properties, namely *next*, *sink* and *cost* denoting the cost to reach a sink node 15 using neighbor 4 as a next hop. Facts can be used to define application-specific data, to trigger reactions and to transparently exchange data with other nodes.

RDL rules are reactive rules, sets of conditional actions with conditions being indicated by a left arrow (<-) and statements, marked by a right arrow (->). The rule *forwarding* will be triggered when both conditions evaluate to true, so at least an instance of a *data_sample* fact as well as a *rt_entry* fact exist in the fact repository, and one of these facts is marked as being an event instance. In this case, a *data_sample* fact is send with a transmission power setting of 100 to the node referenced in the property *next* of the *rt_entry* fact. This integration of a low-level concern such as fine-grained transceiver control into a high-level language mirrors the domain-specific design of RDL.

3.1   Rules: Basic Building Blocks of RDL

RDL rules are neither classical production rules, although their *WHEN condition THEN action* structure may convey this assumption, nor pure ECA rules but integrate characteristics of both worlds into one approach.[1] Instead of providing the denotational semantics of RDL, we will point out how RDL can be positioned in respect to these ends.

---

[1] We'd like to thank the anonymous reviewers for their helpful comments and excellent pointers to related work.

Listing 1.1. Rule for transparent data forwarding

```
1  fact rt_entry [next = 4, sink = 15, cost = 3]
2
3  rule forwarding 100
4  <- exists {data_sample}
5  <- exists {rt_entry}
6  -> send {rt_entry next} 100 {data_sample}
```

Event, condition and action part of RDL rules. Reactive rules generally correspond to change. While production rules trigger upon a modification of their working memory, thus actively maintain rule engine state, ECA rules solely depend on rule-local state, defined by objects and events matched by a rule [2]. Therefore, the semantics of reactivity and its implementation are slightly different: ECA rules, typically denoted in an *ON event IF condition THEN action* format, explicitly name and model events in contrast to a state-driven approach pursued by production rules.

RDL rules are best described as production rules implementing an explicit event concept, a circumstance that can be attributed to its data model. Both events and state are fused into the single data abstraction of a fact, but can be discriminated via information on their state tagged to each individual fact. Formally, $F = \{f_1, f_2, ..., f_n\}, F \subset F^*$ denotes the set of facts stored in the fact repository, where F^* corresponds to the set of all well-formed facts. Each fact has at least a name and may have a set of associated properties. These properties, $P = \{f_1.p_1, f_1.p_2, ..., f_1.p_a, ..., f_n.p_b\}$ within the domain of all well defined properties P^*, $P \subset P^*$, are key-value tuples with available types for values being **bool**, **int**, **string**, **name**. Besides application-specific properties, each fact is additionally tagged with system properties $P = \{f_i.p_{id}, f_i.p_{owner}, f_i.p_{ts}, f_n.p_{state}\}$ including a network-wide unique fact_id, the id of the sensor node that last accessed the fact, a timestamp and a tag reflecting its current processing status. This state flag which can take values {*premodified, modified, unmodified, sticky*} is used on the one hand to mark a fact to be an event instance, in case it is set to *modified*, and on the other to maintain engine-level processing state for subsequent control flow.

To determine whether a rule r_i triggers, the conjunction of its conditions $C_i = \{r_i.c_1, ..., r_i.c_d\}$ has to evaluate to true and one of the facts involved in condition evaluation has to be an event. FACTS supports a push policy for facts, so any sensor reading, any incoming packet or user timer expiration is wrapped into a fact and communicated to the rule engine as a new event. Unlike usual ECA semantics, events are not transient but preserved in the fact repository if not explicitly retracted by a rule, which allows for a straight-forward implementation of predicates on event sequences. Overall, RDL supports test for existence of facts, simple predicates that essentially provide boolean expressions involving fact properties and constants, possibly incorporating basic arithmetic operations, unary predicates such as *min, max, sum* and *count* and allows for their nested composition. Predicates involving temporal order of event occurrences in a distributed, network-level manner can syntactically be expressed, but do need the implementation of a time synchronization protocol, which is not an integral part of the current framework.

Rule evaluation is triggered upon the occurrence of an event. The conditions of an arbitrary set of rules $R = \{r_1, r_2, ..., r_m\}$ with a priority ordering $Prio = \{r_i > r_j \mid$ rule r_i has precedence over $r_j\}$ are then evaluated sequentially according to their priority in a forward chaining manner. Action execution, although possibly composed of a set of different statements $S_i = \{r_i.s_1, ..., r_i.s_e\}$ is always atomic in RDL. Statements for fact definition and retraction, property updates and fact distribution to other wireless sensor nodes are available, as well as a call interface to firmware functionality needed for system interaction.

Rule engine implementation. Whenever events materialized as facts tagged to be modified appear in the fact repository, rule evaluation will take place and only rules that specify conditions matching at least one event facts are eligible to fire. RDL follows a stateful semantic of rule specification, thus allows rules to generate new events other rules may incorporate in their condition part. However, the FACTS implementation adheres a strict ordering of reactions: Any event raised during the evaluation of the rule base cannot trigger rules of the current run. This is preserved by setting these facts to premodified, and finally to modified after completion of the run, scheduling all rules again for evaluation.

The mechanism implemented to pursue forward chaining is a brute-force sequential pattern-matching approach, storing no evaluation state, e.g. in a Rete network at all. Due to the intrinsic memory constraints of the sensor network domain, this is a reasonable decision: The focus is definitely on optimizing memory usage, with neither envisioned sizes of fact repositories, nor those of rule bases requiring optimizations for time consumptions.

3.2   Combination of Interacting Rules: Rulesets

Rule-based implementations can be arbitrarily complex dependent on the functionality to be provided. In some cases, e.g. to realize a threshold-based filtering of certain events according to the attribute values, the implementation is straight-forward: for every possible state that can be reached after the event emerged, a rule is provided to specify the necessary guards on state transition in the condition part and encapsulate the intended action in the statement part. For more advanced algorithms, e.g. the implementation of a routing protocol or a sophisticated aggregation scheme, more elaborate mechanisms such as the usage of production rules are usually required to obtain the desired functionality. The focus then moves from single rules to sets of interacting rules, which naturally implies a careful development to ensure the termination of rule evaluation. From a developers perspective, it is desirable to reflect this semantical unity of sets of rules with a syntactical equivalent, thus enable an encapsulation of concerns. RDL therefore offers to combine rules into named sets of rules or *rulesets* and provides means to ensure their evaluation order. The advantages of providing distinct rulesets instead of monolithic rule-bases are obvious: the added modularity improves maintainability and encourages reuse. Functionality shared by various applications can be swapped e.g. into dedicated middleware rulesets and linked on demand. Furthermore, from a sensor network perspective, over-the-air code dissemination is then much more energy efficient, since partial code updates are possible.

With the advantages at hand, the need for several additions to RDL become apparent. Interface specification is problematic when rule evaluation relies on matching of fact names, scoping has to be addressed and namespaces added to prevent faulty system behavior.

Relational References for Generic Matching. Rules trigger in case facts match the specified conditions. Recall the example in Listing 1.1, where a rule implements a forwarding mechanism for instances of the event *data_sample*. Since rule evaluation and action execution depend on matching and no means for dynamic binding of variables is available at runtime, this rule will solely forward facts with the name data_sample. As a consequence, we need to specify a rule for each event instance occurring in the system that is to be forwarded, a circumstance leading to bloated code. If we compare this situation to other programming languages, the mechanism we're missing is polymorphism, thus allowing facts of different types to be handled using a uniform interface. The scope of demand for transparent reuse can be a single rule, as well as a complete ruleset.

We objected to this problem by extending the possible data types for fact properties and added *name* to available types as a means to explicitly reference event classes. A fact can then incorporate a property that refers to other facts, whose content will be fetched when indicated by an asterisk (*). Note however that although one is tempted to put this on the same level as a pointer e.g. in C, the semantics are different: While a C pointer points to exactly one location in memory, a property of type *name* is an indirection to possibly multiple matching facts. Listing 1.2 serves as a simplified example to clarify this and to demonstrate the syntax for referring to referenced facts.

The main goal has been to modify the rule *forwarding* in a way that allows it to be applied to facts of various types. Therefore, instead of solely matching a dedicated fact to evaluate rule conditions, the condition in line 4 specifies the rule engine to test whether a fact *data_sample*, as well as a fact that the property *sensor* refers to reside in the fact repository. The second rule provides such a fact: In case a new *humidity* reading is pushed into the fact repository, a *data_sample* fact is created that points to humidity facts via the attribute *sensor*. The first rule will transparently forward every fact matching the name specified in the facts *sensor* property. This way, a third rule e.g. reacting on a new temperature sample can as well trigger temperature forwarding via the provided interface. Naturally, this concept can be adopted across ruleset boundaries, given the fact that access to interfaces, represented by fact names, is not restricted, see 3.2.

An alternative could have been to implement superclasses for facts or events respectively, e.g. to specify all sensor readings to be instances of a superclass *data_sample* and enable transparent matching of instances against superclass

Listing 1.2. Polymorphic implementation for transparent forwarding

```
 1  ruleset TransparentForwarding
 2
 3  public name data_sample = "data_sample"
 4
 5  rule forwarding 100
 6  <- exists *{data_sample sensor}
 7  <- exists {routing_entry}
 8  -> send {routing_entry next_hop} 100   *{data_sample
        sensor}
 9
10  rule sensor_humidity 99
11  <- exists {humidity}
12  -> define data_sample [sensor = humidity]
```

rules. Although this solution of utilizing inheritance is syntactically more appealing, the drawbacks are evident: To enable access to multiple interfaces, all corresponding classes have to be inherited and provided functionality can not be applied on demand but is strictly executed, even if not intended.

Namespaces and Scoping. Since facts serve as a means of internal communication of system state between rules for forward chaining, the probability of an unintentional clash of names when sharing a global namespace among rulesets increases with the number of rules provided. A simple, yet effective mechanism to avoid this, is to enable the declaration of namespaces, which equal to the scope of a ruleset, and restricted scope of names accordingly.

We adopt the widely used and well-understood keywords *private* and *public* in RDL to enable clear interface definitions via fact names. For instance, the name *data_sample* in the above example is declared to be public (line 1). Therefore, any ruleset can benefit from utilizing the forwarding routine, but at the same time is requested to understand the impact of specifying such a fact in its own scope has. Compile-time checking of ruleset interdependencies and correct fact references alleviate the development process and support less error-prone implementations.

4   Related Work

The need to address reactivity in system design has seen years of research in a variety of domains, including active databases, SCADA systems, business process management or in the area of semantic web reasoning. The general idea of rule-based language utilization in the wireless sensor network community has always been present, see e.g. [3], but actual specifications of rule languages [4] and available implementations of rule based systems are rare. FACTS relies on such a language abstraction and integrates concepts from very different areas including tuple spaces, event-driven architectures and production rule systems [2] into a dedicated, domain-specific approach. Work in this area that shares a lot of similarities with ours is e.g. TeenyLime [5], an implementation that offers transparent sharing of one-hop tuple spaces and registration of reactions to emerging tuples. While TeenyLime is available to a programmer as an API to overcome distribution issues, the focus of FACTS with its runtime environment is clearly on providing a suitable abstraction from embedded, event-driven programming.

5   Conclusions

Reactive rules are a powerful concept worth porting to embedded devices. In this paper we reviewed general concepts of our language RDL developed specifically for wireless sensor networks and examined its semantics in respect to ECA and production rules. Furthermore, we presented extensions to the language that allow for a better separation of concerns when utilizing forward chaining rules by adding mechanisms for scoping and polymorphic matching. The increase in modularity these extensions grant will not only be beneficial regarding the software development process due to facilitated code reuse, but provide an important building block for energy-efficient, network-wide code updates at runtime.

References

1. Terfloth, K., Wittenburg, G., Schiller, J.: FACTS - A Rule-Based Middleware Architecture for Wireless Sensor Networks. In: Proceedings of the First International Conference on COMmunication System softWAre and MiddlewaRE (COMSWARE 2006), New Delhi, India (January 2006)
2. Berstel, B., Bonnard, P., Bry, F., Eckert, M., Patranjan, P.L.: Reactive rules on the web. In: Antoniou, G., Aßmann, U., Baroglio, C., Decker, S., Henze, N., Patranjan, P.-L., Tolksdorf, R. (eds.) Reasoning Web. LNCS, vol. 4636, pp. 183–239. Springer, Heidelberg (2007)
3. Kasten, O., Römer, K.: Beyond event handlers: Programming wireless sensors with attributed state machines. In: The Fourth International Conference on Information Processing in Sensor Networks (IPSN), Los Angeles, USA, pp. 45–52 (April 2005)
4. Sen, S., Cardell-Oliver, R.: A rule-based language for programming wireless sensor actuator networks using frequency and communication. In: Proceedings of the 3rd Workshop on Embedded Networked Sensors (EmNets), Cambridge, USA (2006)
5. Costa, P., Mottola, L., Murphy, A.L., Picco, G.P.: Programming wireless sensor networks with the teenylime middleware. In: Proceedings of the 8th ACM/IFIP/USENIX International Middleware Conference (Middleware 2007), Newport Beach, CA, USA (November 2007)

Local and Distributed Defeasible Reasoning in Multi-Context Systems

Antonis Bikakis and Grigoris Antoniou

Institute of Computer Science, FO.R.T.H., Vassilika Voutwn
P.O. Box 1385, GR 71110, Heraklion, Greece
{bikakis,antoniou}@ics.forth.gr

Abstract. Multi-Context Systems (MCS) are logical formalizations of distributed context theories connected through a set of mapping rules, which enable information flow between different contexts. Reasoning in MCS introduces many challenges that arise from the heterogeneity of contexts with respect to the language and inference system that they use, and from the potential conflicts that may arise from the interaction of context theories through the mappings. This study proposes a P2P rule-based reasoning model for MCS, which handles (a) incomplete or inconsistent local context information, by representing contexts as local theories of Defeasible Logic and performing local defeasible reasoning, and (b) global inconsistencies that result from the integration of local contexts, by representing mappings as defeasible rules and performing some type of distributed defeasible reasoning. It also provides a distributed algorithm for query evaluation, analyzes its formal properties, and illustrates its use in a Semantic Web use case scenario.

1   Motivation and Background

A Multi-Context System consists of a set of *contexts* and a set of inference rules (known as *mapping* or *bridge* rules) that enable information flow between different contexts. A context can be thought of as a logical theory - a set of axioms and inference rules - that models local context knowledge. Different contexts are expected to use different languages and inference systems, and although each context may be locally consistent, global consistency cannot be required or guaranteed. Reasoning with multiple contexts requires performing two types of reasoning; (a) *local reasoning*, based on the individual context theories; and (b) *distributed reasoning*, which combines the consequences of local theories using the mappings. The most critical issues of contextual reasoning are; (a) the *heterogeneity* of local contexts (with respect to the language and inference system that they use); and (b) the potential conflicts that may arise both in local contexts, and also from the interaction of different contexts through the mappings. Our study focuses on the second issue, by modeling contexts as local non-monotonic rule theories in a P2P system, and performing some type of defeasible reasoning on the distributed context theories.

N. Bassiliades, G. Governatori, and A. Paschke (Eds.): RuleML 2008, LNCS 5321, pp. 135–149, 2008.

The notions of *context* and *contextual reasoning* were first introduced in AI by McCarthy in [1], as an approach for the problem of *generality*. In the same paper, he argued that the combination of non-monotonic reasoning and contextual reasoning would constitute an adequate solution to this problem. Since then, two main formalizations have been proposed to formalize context: the propositional logic of context (*PLC* [2,3]) , and the Multi-Context Systems introduced in [4], which later became associated with the Local Model Semantics proposed in [5]. The second formalism was the basis of two recent studies that were the first to deploy non-monotonic reasoning approaches in MCS: (a) the non-monotonic rule-based MCS framework, which supports default negation in the mapping rules allowing to reason based on the absence of context information, proposed in [6]; and (b) the multi-context variant of Default Logic [7]. The latter models the bridge relations between different contexts as *default rules*, and has the additional advantage that is closer to implementation due to the well-studied relation between Default Logic and Logic Programming. However, the authors do not provide specific reasoning algorithms (e.g. for query evaluation), and their model does include the notion of priority, which we use for conflict resolution.

Our study also relates to several recent studies that are focused on formal models and methods for reasoning in peer data management systems. A key issue in formalizing data-oriented P2P systems is the semantic characterization of *mappings* (bridge rules). One approach (followed in [8,9]) is the first-order logic interpretation of P2P systems. [10] identified several drawbacks with this approach, regarding modularity, generality and decidability, and proposed new semantics based on epistemic logic. A common problem of both approaches is that they do not model and thus cannot handle inconsistency. Franconi *et al.* in [11] extended the autoepistemic semantics to formalize local inconsistency. The latter approach guarantees that a locally inconsistent database base will not render the entire knowledge base inconsistent. A broader extension, proposed by Calvanese *et al.* in [12], is based on non-monotonic epistemic logic, and enables isolating local inconsistency, while also handling peers that may provide mutually inconsistent data. The proposed query evaluation algorithm assumes that all peers share a common alphabet of constants, and does not model *trust* or *priorities* between the peers. The propositional P2P inference system proposed by Chatalic *et al.* in [13] deals with conflicts caused by mutually inconsistent information sources, by detecting them and reasoning without them. The main problem is the same, once again: To perform reasoning, the conflicts are not actually resolved using some external trust or priority information; they are rather isolated.

This study proposes a reasoning model that represents contexts as local rule theories in a P2P system. In order to support cases of incomplete or inconsistent local information, we represent contexts as theories of Defeasible Logic. These theories include strict (monotonic) rules that express definite (sound) knowledge, and defeasible rules, which are used to express uncertainty about their conclusions. In such theories, local rules may support contradictory conclusions. In the case that two competing rules can be applied (their premises are consequences of the local theory), we use the rule priority relation of the local theory to resolve the conflict. This relation gives priority to one of the two competing rules, while blocking the other.

Even if all context theories are locally consistent, we cannot assume consistency in the global knowledge base. The unification of local theories may result in inconsistencies caused by the mappings. For example, a context theory A may import context knowledge from two different contexts B and C, through two competing mapping rules. In this case, even if the three different contexts are locally consistent, their unification through the mappings defined by A may contain inconsistencies. To deal with this type of inconsistencies (*global conflicts*), we model mappings as defeasible rules and use additional preference information, which e.g. may express trust information in the different contexts, to resolve them. So, if context A trusts B more than C, it will give higher priority to the mappings that import context knowledge from B.

With this model, we aim to capture the three fundamental dimensions of contextual reasoning, as these were formulated in [14]; namely *partiality, approximation* and *perspective*.

- *Partiality.* Each peer may not have immediate access to all available information, so a peer theory can be thought as a partial representation of *the world*.
- *Approximation* Each peer theory differs at the level of detail at which a portion of *the world* is represented.
- *Perspective* Each peer theory encodes a different point of view on *the world*.

Furthermore, the P2P paradigm enables us to model:

- Information flow between different contexts as message exchange between the system peers.
- Context changes using the dynamics of a P2P system.
- Confidence in the different context theories as trust between the system peers.

The rest of this paper is structured as follows: In the next section, we describe the proposed reasoning model. In section 3, we provide a specific reasoning algorithm for distributed query evaluation, and in Section 4, we analyze its formal properties. In section 5, we illustrate how the algorithm can be applied in a scenario from the Semantic Web domain. In the last section, we summarize and refer to the next steps of this work.

2   Reasoning Model

Our approach models a Multi-Context System P as a collection of distributed local rule theories P_i in a P2P system:

$$P = \{P_i\}, i = 1, 2, ..., n$$

Each system peer (context) has a proper distinct vocabulary V_i and a unique identifier i. Each local theory is a set of rules that contain only local literals (literals from the local vocabulary). There are two types of local rules:

- Strict rules, of the form

$$r_i^l : a_i^1, a_i^2, ... a_i^{n-1} \rightarrow a_i^n$$

  where i denotes the peer identifier. These rules express strict (sound) knowledge and are interpreted in the classical sense: whenever the literals in the body of a strict rule $(a_i^1, a_i^2, ... a_i^{n-1})$ are strict consequences of the theory, then so is the conclusion of the rule (a_i^n). Strict rules with empty body are used to express factual knowledge.
- Defeasible rules, of the form

$$r_i^d : b_i^1, b_i^2, ... b_i^{n-1} \Rightarrow b_i^n$$

  Defeasible rules are used to express uncertainty, in the sense that a rule of this type (r_i^d) cannot be applied to support its conclusion (b_i^n) if there is adequate contrary evidence.

Each peer also defines mappings that associate literals from its own vocabulary (*local literals*) with literals from the vocabulary of other peers (*foreign literals*). The acquaintances of peer P_i, $ACQ(P_i)$ are the set of peers that at least one of P_i's mappings involves at least one of their local literals. The mappings are also modeled as defeasible rules of the form:

$$r_i^m : a_i^1, a_j^2, ... a_k^{n-1} \Rightarrow a_i^n$$

The above mapping rule is defined by P_i, and associates some of its own local literals with some of the local literals of P_j, P_k and other system peers. a_i^n is a local literal of the theory that has defined r_i^m.

A peer also defines an acyclic priority relation $>$ on the set of its local and mapping rules (that is, the transitive closure of $>$ is irreflexive). This is of the form $r_i > r_i'$ and is used to resolve the conflict that arises in case the premises of both rules are consequences of the local and mapping rules of the peer theory; in this case, r_i overrides r_i', and can be applied to support its conclusion, while r_i' is blocked.

Finally, each peer P_i defines a trust level order T_i, which includes a subset of the system peers, and expresses the trust that P_i has in the other system peers. This is of the form:

$$T_i = [P_k, P_l, ..., P_n]$$

The peers that are not included in T_i are less trusted by P_i than those that are part of the list.

In the reasoning model that we described, two types of conflicts may arise: (a) *local conflicts* between competing local rules; in this case, we use the rule priority relation of the local theory to resolve the conflicts; (b) *global conflicts*, which derive from the unification of the local theories through the mappings; to resolve this type of conflicts, we may have to use both rule priorities and the trust level orderings of the system peers.

3 The $P2P\_DR_{dl}$ Algorithm

$P2P\_DR_{dl}$ is a distributed algorithm for query evaluation in Multi-Context Systems following the model that we described in the previous section. The specific reasoning problem that it deals with is: *Given a MCS P, and a query about literal x_i issued to peer P_i, find the truth value of x_i considering P_i's local theory, its mappings and the context theories of the other system peers.* The algorithm parameters are:

x_i: the queried literal
P_0: the peer that issues the query
P_i: the local peer
SS_{x_i}: the Supportive Set of x_i (a set of literals that is initially empty)
CS_{x_i}: the Conflicting Set of x_i (a set of literals that is initially empty)
$Hist_{x_i}$: the list of pending queries ($[x_1, ..., x_i]$)
Ans_{x_i}: the answer returned for x_i (initially empty)

The algorithm proceeds in four main steps. In the first step (lines 1-10 in the pseudocode given below), the algorithm determines if the queried literal, x_i, or its negation $\neg x_i$ are consequences of P_i's local strict rules. To do that it calls a local reasoning algorithm (*local_alg*, described later in this section), which returns a positive answer, in case x_i derives from the local strict rules, or a negative answer in any other case. In the code below, we denote as $R_s(x_i)$ the set of (local or mapping) rules that support x_i (as their conclusion); and as $R_c(x_i)$, the set of rules that contradict x_i (those that support $\neg x_i$).

If Step 1 fails, the algorithm collects, in the second step (lines 11-30), the local and mapping rules that support x_i. To check which of these rules can be applied, it checks the truth value of the literals in their body by issuing similar queries (recursive calls of the algorithm) to P_i or to the appropriate neighboring peers $P_j \in ACQ_{P_i}$ (line 19). To avoid cycles, before each new query, it checks if the same query has been issued before, during the same algorithm call (using $Hist$). For each applicable supportive rule r_i, the algorithm builds its supportive set SS_{r_i}; this derives from the unification of the set of the *foreign literals* (literals that are defined by peers that belong in $ACQ(P_i)$) that are contained in the body of r_i, with the Supportive Sets of the local literals that belong in the body of the same rule (lines 22-25). In the end, in case there is no applicable supportive rule ($SR_{x_i} = \{\}$, where SR_{x_i} is the set of applicable rules that support x_i), the algorithm returns a negative answer for x_i and terminates. Otherwise, it computes the Supportive Set of x_i, SS_{x_i}, as the *strongest* of the Supportive Sets of the applicable rules that support x_i, and proceeds to the next step. The *strongest* Supportive Set is computed using the *Stronger* function (described later in this section), which applies the preference relation defined by P_i, T_i, on the given sets.

In the third step (lines 31-50), in the same way with the previous step, the algorithm collects the rules that contradict x_i and builds the conflicting set of x_i (CS_{x_i}). In case there is no applicable rule that contradicts x_i, the algorithm terminates by returning a positive answer for x_i. Otherwise, it proceeds with the last step. In the code below, we denote as CR_{x_i} the set of the applicable rules that contradict (support the negation of) x_i.

In the last step (lines 51-54), $P2P\_DR_{dl}$ determines the truth value of x_i considering the priority relation in P_i and the trust level ordering, T_i. If for each rule that can be applied to contradict x_i ($r' \in CR_{x_i}$) there is a superior (according to the priority relation), or a non-inferior but stronger (according to T_i) applicable supportive rule ($r \in SR_{x_i}$), the algorithm returns a positive truth value. In any other case (including the case that there is not enough priority or trust information available), it returns a negative answer. To compare two conflicting rules based on T_i, the algorithm uses the *Stronger* function. Below, we give the code of the algorithm.

$P2P\_DR_{dl}(x_i, P_0, P_i, SS_{x_i}, CS_{x_i}, Hist_{x_i}, Ans_{x_i}, T_i)$

```
1.  if ∃r_i^l ∈ R_s(x_i) then
2.      localHist_{x_i} ← [x_i]
3.      call local_alg(x_i, localHist_{x_i}, localAns_{x_i})
4.      if localAns_{x_i} = Yes then
5.          return Ans_{x_i} = localAns_{x_i} and terminate
6.  if ∃r_i^l ∈ R_c(x_i) then
7.      localHist_{x_i} ← [x_i]
8.      call local_alg(¬x_i, localHist_{x_i}, localAns_{¬x_i})
9.      if localAns_{¬x_i} = Yes then
10.         return Ans_{x_i} = ¬localAns_{¬x_i} and terminate
11. SR_{x_i} ← {}
12. for all r_i^{ldm} ∈ R_s(x_i) do
13.     SS_{r_i} ← {}
14.     for all b_t ∈ body(r_i^{ldm}) do
15.         if b_t ∈ Hist_{x_i} then
16.             stop and check the next rule
17.         else
18.             Hist_{b_t} ← Hist_{x_i} ∪ b_t
19.             call P2P_DR_{dl}(b_t, P_i, P_t, SS_{b_t}, CS_{b_t}, Hist_{b_t}, Ans_{b_t}, T_t)
20.             if Ans_{b_t} = No then
21.                 stop and check the next rule
22.             else if Ans_{b_t} = Yes and b_t ∉ V_i then
23.                 SS_{r_i} ← SS_{r_i} ∪ b_t
24.             else
25.                 SS_{r_i} ← SS_{r_i} ∪ SS_{b_t}
26.         if SR_{x_i} = {} or Stronger(SS_{r_i}, SS_{x_i}, T_i) = SS_{r_i} then
27.             SS_{x_i} ← SS_{r_i}
28.             SR_{x_i} ← SR_{x_i} ∪ r_i^{ldm}
29. if SR_{x_i} = {} then
30.     return Ans_{x_i} = No and terminate
31. CR_{x_i} ← {}
32. for all r_i^{ldm} ∈ R_c(x_i) do
33.     SS_{r_i} ← {}
34.     for all b_t ∈ body(r_i^{ldm}) do
```

```
35.        if b_t ∈ Hist_{x_i} then
36.            stop and check the next rule
37.        else
38.            Hist_{b_t} ← Hist_{x_i} ∪ b_t
39.            call P2P_DR_{dl}(b_t, P_i, P_t, SS_{b_t}, CS_{b_t}, Hist_{b_t}, Ans_{b_t}, T_t)
40.            if Ans_{b_t} = No then
41.                stop and check the next rule
42.            else if Ans_{b_t} = Yes and b_t ∉ V_i then
43.                SS_{r_i} ← SS_{r_i} ∪ b_t
44.            else
45.                SS_{r_i} ← SS_{r_i} ∪ SS_{b_t}
46.    if CR_{x_i} = {} or Stronger(SS_{r_i}, CS_{x_i}, T_i) = SS_{r_i} then
47.        CS_{x_i} ← SS_{r_i}
48.    CR_{x_i} ← CR_{x_i} ∪ r_i^{ldm}
49. if CR_{x_i} = {} then
50.    return  Ans_{x_i} = Yes and SS_{x_i} and terminate
51. for all r_i' ∈ CR_{x_i} do
52.    if ∄r_i ∈ SR_{x_i}: r_i > r_i' or
        (r_i' ≯ r_i and Stronger(SS_{r_i}, SS_{r_i'}, T_i) = SS_{r_i}) then
53.        return  Ans_{x_i} = No and terminate
54. return  Ans_{x_i} = Yes and SS_{x_i}
```

The local reasoning algorithm *local_alg* is called by *P2P_DR* to determine if a literal is a consequence of the strict local rules of the theory. The algorithm parameters are:

x_i: the queried literal
$localHist_{x_i}$: the list of pending queries in P_i
$localAns_{x_i}$: the local answer for x_i (initially No)

local_alg$(x_i, localHist_{x_i}, localAns_{x_i})$

```
 1. for all r_i^l ∈ R_s(x_i) do
 2.    if body(r_i^l) = {} then
 3.        return  localAns_{x_i} = Yes and terminate
 4.    else
 5.        for all b_i ∈ body(r_i^l) do
 6.            if b_i ∈ localHist_{x_i} then
 7.                stop and check the next rule
 8.            else
 9.                localHist_{b_i} ← localHist_{x_i} ∪ b_i
10.                call local_alg(b_i, localHist_{b_i}, localAns_{b_i})
11.        if for every b_i: localAns_{b_i} = Yes then
12.            return  localAns_{x_i} = Yes and terminate
```

The *Stronger(S, C, T)* function is used by *P2P_DR* to check which of S and C sets is *stronger*, based on T (the preference relation defined by the peer that the algorithm is called by). According to T, a literal a_k is considered to be *stronger* than a_l if P_k precedes P_l in T. The strength of a set is determined by the the weakest literal in this set.

Stronger(S, C, T)

1. $a^w \leftarrow a_k \in S$ s.t. *for all* $a_i \in S : P_k$ does not precede P_i in T
2. $b^w \leftarrow a_l \in C$ s.t. *for all* $b_j \in C : P_l$ does not precede P_j in T
3. **if** P_k precedes P_l in T **then**
4. $Stronger = S$
5. **else if** P_l precedes P_k in T **then**
6. $Stronger = C$
7. **else**
8. $Stronger = None$

Below we demonstrate how the algorithm works through an example. In the MCS depicted in Figure 1, there are four peer theories, and a query about x_1 is issued to P_1. Notice that in P_2 there is a local conflict caused by the competing rules r_{23}^d and r_{24}^d, and that the interaction of P_1 with other peer theories through mapping rules r_{12}^m and r_{13}^m cause a global conflict about the truth value of a_1.

$$
\begin{array}{lll}
\underline{P_1} & \underline{P_2} & \underline{P_3} \\
r_{11}^l : a_1 \to x_1 & r_{21}^l :\to b_2 & r_{31}^l :\to a_3 \\
r_{12}^m : a_2 \Rightarrow a_1 & r_{22}^l :\to c_2 & \\
r_{13}^m : a_3, a_4 \Rightarrow \neg a_1 & r_{23}^d : b_2 \Rightarrow a_2 & \underline{P_4} \\
 & r_{24}^d : c_2 \Rightarrow \neg a_2 & r_{41}^l :\to a_4 \\
 & r_{23}^d > r_{24}^d &
\end{array}
$$

Fig. 1. A MCS of Four Context Theories

- In the first step, the algorithm fails to determine the truth value of x_1 using P_1's strict rules.
- It successively calls rules r_{11}^l, r_{12}^m, and issues a query about a_2.
- In P_2, a_2 does not derive from the local strict rules, so the algorithm checks the applicability of the only rule that supports a_2, r_{23}^d, by issuing a query about b_2, which is its only premise.
- b_2 derives in P_2 as a consequence of r_{21}^l, so the algorithm determines that r_{23}^d is applicable.
- In a similar way, it determines that the only rule that contradicts a_2, r_{23}^d, is also applicable, as its only premise, c_2 is locally proved.
- Using the priority relation $r_{23}^d > r_{24}^d$, it resolves the local conflict, and computes a positive answer for a_2.
- It determines that r_{12}^m is applicable, and builds its Supportive Set, which contains literal a_2 ($SS_{r_{12}^m} = \{a_2\}$).
- Using a similar process, the algorithm determines that r_{13}^m (the only rule that contradicts a_1) is also applicable, and computes its Supportive Set $SS_{r_{13}^m} = \{a_3, a_4\}$.
- As there is no priority between the two conflicting rules (r_{12}^m and r_{13}^m), the algorithm uses the trust level order defined by P_1, T_1, to compare their supportive sets. Assuming that $T_1 = [P_4, P_2, P_3]$, a_2 and a_3 are respectively the weakest elements of $SS_{r_{12}^m}$ and $SS_{r_{13}^m}$, and a_3 is weaker than a_2; so r_{12}^m is

stronger than r_{13}^m. Consequently, it computes a positive answer for a_1, and determines that r_{11}^l is applicable.

- As there is no rule that contradicts x_1, it eventually returns a positive answer for x_1.

4   Properties of $P2P\_DR_{dl}$

In this section we describe some formal properties of $P2P\_DR_{dl}$ with respect to its termination (Proposition 1), complexity (Propositions 2-3), and the possibility to create an equivalent unified defeasible theory from the distributed context theories (Theorem 1). The proofs for Propositions 1-3 and Theorem 1 are available in [15]. Proposition 1 holds as cycles are detected within the algorithm.

Proposition 1. *The algorithm is guaranteed to terminate returning either a positive or a negative answer for the queried literal.*

Prop. 2 is a consequence of two states that we retain for each peer, which keep track of the incoming and outgoing queries of the peer.

Proposition 2. *The total number of messages that are exchanged between the system peers for the computation of a single query is $O(n^2)$ (in the worst case that all peers have defined mappings with all the other system peers), where n stands for the total number of system peers.*

Proposition 3. *The computational complexity of the algorithm on a single peer is in the worst case $O(n^2 \times n_l^2 \times n_r + n \times n_l \times n_r^2)$, where n stands for the total number of system peers, n_l stands for the number of literals a peer may define, and n_r stands for the total number of (local and mapping) rules that a peer theory may contain.*

4.1   Equivalent Unified Defeasible Theory

The goal of the procedure that we describe below is to build a global defeasible theory $T_v(P)$, which produces the same results as the application of $P2P\_DR_{dl}$ on MCS P. The existence of such theory enables us to resort to centralized reasoning in cases that there is need for central control. The procedure consists of the following steps:

1. The local strict rules of each peer theory are added as strict rules in $T_v(P)$.
2. The local defeasible and mapping rules of each peer theory are added as defeasible rules in $T_v(P)$.
3. The priority relations on all pairs of conflicting rules of each peer theory are added as rule priorities on the corresponding pairs of rules in $T_v(P)$.
4. For each pair of conflicting rules, for which there is no priority information in $T_v(P)$, we add a priority relation using the $Priorities_{dl}$ process that we describe below.

The role of *Priorities$_{dl}$* is to augment $T_v(P)$, as it derives from the first three steps that we describe above, with the additional required rule priorities considering the trust level orders of the system peers. In this process we use two special elements: (a) w, to mark the rules that cannot be applied; and (b) s, to mark the literals the truth value of which derives from the strict rules of $T_v(P)$. The process takes as input a literal of the theory, say x_i, the strict and defeasible rules of $T_v(P)$ that support or contradict x_i ($R[x_i]$, $R[\neg x_i]$), and the trust ordering of P_i, T_i, and returns the supportive set of x_i (S_{x_i}), and augments $T_v(P)$ with the required priority relations. The algorithm follows three main steps:

In the first step (lines 1-12 in the pseudocode given below), it builds the Supportive Sets for the rules that support or contradict x_i. These sets are built in a similar way that *P2P_DR$_{dl}$* computes the Supportive Sets of the respective rules in the distributed theories, with only one difference: If there is a literal in the body of a rule that contains w in its Supportive Set, then the algorithm assigns $\{w\}$ as the Supportive Set of the rule, meaning that this rule is inapplicable.

In the second step (lines 13-17), *Priorities$_{dl}$* collects all the pairs of applicable conflicting rules, for which there is no priority relation, and adds suitable priority relations by applying the *Stronger* function on their Supportive Sets (using the trust level order of the peer that defined x_i, T_i).

In the final step (lines 18-31), the algorithm computes the Supportive Set of x_i using the following rules: (a) If there is no applicable supportive rule, it returns $\{w\}$ (lines 18-19); (b) If there is a strict rule, the premises of which are supported only by applicable strict rules, it returns $\{s\}$ (lines 20-21); (c) If there is an applicable contradicting rule that is not inferior to any applicable supportive rule, it returns $\{w\}$ (lines 22-30); and (d) In any other case it returns the Supportive Set of the strongest applicable rule, using the *Stronger* function and T_i (line 31). The code of the process is given below.

Priorities$_{dl}$(x_i,$R[x_i]$,$R[\neg x_i]$, T_i, S_{x_i})

```
1.  for all r_i ∈ R[x_i] ∪ R[¬x_i] do
2.      S_{r_i} ← {}
3.      for all a_i ∈ body(r_i) ∩ V_i do
4.          call Priorities_{dl}(a_i,R[a_i],R[¬a_i], T_i, S_{a_i})
5.          S_{r_i} ← S_{r_i} ∪ S_{a_i}
6.      for all a_j ∈ body(r_i) \ V_i do
7.          call Priorities_{dl}(a_j,R[a_j],R[¬a_j], T_j, S_{a_j})
8.          if w ∈ S_{a_j} then
9.              S_{r_i} ← {w}
10.             stop and check next r_i
11.         else
12.             S_{r_i} ← S_{r_i} ∪ a_j
13. for all pairs (r_i ∈ R[x_i], s_i ∈ R[¬x_i])| w ∉ S_{r_i}, w ∉ S_{s_i}, r_i > s_i, s_i < r_i ∉ T_v(P)
    do
14.     if Stronger(S_{r_i}, S_{s_i}, T_i) = S_{r_i} then
15.         add r_i > s_i in T_v(P)
16.     else if Stronger(S_{r_i}, S_{s_i}, T_i) = S_{s_i} then
```

17.   add $s_i > r_i$ in $T_v(P)$
18. **if** $\nexists r_i \in R[x_i] | w \notin S_{r_i}$ **then**
19.   **return** $S_{x_i} = \{w\}$ and terminate
20. **if** $\exists r_i \in R[x_i] |$ for all $a_i \in body(r_i)$: $S_{a_i} = \{s\}$ **then**
21.   **return** $S_{x_i} = \{s\}$ and terminate
22. **for all** $s_i \in R[\neg x_i]$ **do**
23.   **if** $w \in S_{s_i}$ **then**
24.   stop and check next s_i
25.   **else**
26.   **for all** $r_i \in R[x_i] | w \notin S_{r_i}$ **do**
27.   **if** $r_i > s_i \in T_v(P)$ or $(s_i > r_i \notin T_v(P)$ and $Stronger(S_{r_i}, S_{s_i}, T_i) = S_{r_i})$
              then
28.   stop and check next s_i
29.   **else**
30.   **return** $S_{x_i} = \{w\}$ and terminate
31. **return** $S_{x_i} = S_{r_i} | r_i \in R[x_i]$ and $w \notin S_{r_i}$ and for all $t_i \in R[x_i] \setminus r_i | w \notin S_{t_i}$:
     $Stronger(S_{r_i}, S_{t_i}, T_i) \neq S_{t_i}$

To prove Theorem 1, which follows, we use the following definition:

Definition 1. *A MCS P is* **acyclic** *iff there is no rule* $r \in P$ *such that the conclusion of r may be used to prove a literal in the body of r.*

Theorem 1. *The global defeasible theory* $T_v(P)$, *augmented with the priority relations derived from the application of* $Priorities_{dl}$ *on all literals of the theory, produces, under the proof theory of [16], the same results as the application of* $P2P\_DR_{dl}$ *on the distributed context theories of an acyclic MCS P.*

The latter property, which shows the equivalence with a defeasible theory, enables resorting to centralized reasoning by collecting the distributed context theories in a central entity and creating an equivalent defeasible theory. The complexity of $Priorities_{dl}$ used for the derivation of the equivalent global theory is comparable with the complexity of $P2P\_DR_{dl}$. Via Theorem 1, $P2P\_DR_{dl}$ also has a precise semantic characterization. Defeasible Logic has a proof-theoretic [16], an argumentation-based [17] and a model-theoretic semantics [18].

5   Application in the Semantic Web Domain

In this section, we illustrate how $P2P\_DR_{dl}$ works in a use case scenario from the Semantic Web domain, where the MCS model that we described in Section 2 is used to represent local knowledge and interaction between cooperating agents in a web environment. The mapping rules, which allow information exchange between the agents, are actually ontology mappings, and $P2P\_DR_{dl}$ is used to resolve any conflicts that may be caused by mutually inconsistent mappings.

In this scenario, we consider a web agent that has been configured by Adam (a student of Computer Science) to crawl the web, seek for articles that fit Adam's preferences, and recommend these articles to Adam. For each article, the crawler gathers specific information, e.g. the name of the journal that contains the article, and the article keywords. In the next step, the agent matches the article data with the premises of the rules that express Adam's preferences. In case, the agent can reach to a decision based on its strict local rules, it terminates its job by recommending / not recommending the article to the student. Otherwise, it uses the university P2P network trying to contact similar agents operating in the same web environment on behalf of Adam's fellow-students. We assume that Adam has predefined mappings between his vocabulary and the vocabulary used by some of his fellow-students. Using these mappings, Adam's agent contacts a suitable set of similar agents, which may help it decide about the article, using their local knowledge.

Consider that Adam's agent finds a new article, that has been recently published in the 'AmI Journal' (an imaginary journal), and has three keywords; 'contextual reasoning', 'defeasible logic', 'blackboard system'. The agent expresses these facts in terms of local strict rules with empty body.

$\rightarrow ami\_journal$
$\rightarrow contextual\_reasoning$
$\rightarrow defeasible\_logic$
$\rightarrow blackboard\_system$

Consider also that Adam has the following preferences: 'Articles about Ubiquitous Computing are not interesting, unless they deal with distributed and non-monotonic methods for reasoning about context.' These preferences are expressed in Adam's agent as follows:

$r^d_{A1} : ubiquitous\_computing \Rightarrow \neg recommend$
$r^d_{A2} : nonmonotonic\_reasoning, distributed\_reasoning,$
$contextual\_reasoning \Rightarrow recommend$
$r^d_{A2} > r^d_{A1}$

Adam's local knowledge does not contain any information about how *ami journal*, *defeasible_logic* and *blackboard_system* are related to the literals used in the bodies of r^d_{11} and r^d_{12}. Therefore, the agent cannot determine about the recommendation of the article based on its local knowledge, and proceeds to the next step, in which it uses its mapping rules. Adam has predefined four mapping rules $(r^m_{A3}\text{-}r^m_{A6})$, which associate part of its vocabulary with the vocabularies used by the agents of his fellow-students Bill, Christine, Daniel and Ellen. Bill's research interests focus on ambient computing. Christine is interested in logics, while Daniel and Ellen are both interested in distributed architectures. Rule r^m_{A3} associates *ambient_computing* defined by Bill with the concept of *ubiquitous_computing*. r^m_{A4} connects the concepts of *nonmonotonic_logic* (used by Christine) and *nonmonotonic_reasoning*. r^m_{A5} states that *centralized_architecture* defined by Daniel implies non-*distributed_reasoning*, while r^m_{A6} expresses that *decentralized_reasoning* (defined by Ellen) is a type of *distributed_reasoning*.

$r^m_{A3} : ambient\_computing_B \Rightarrow ubiquitous\_computing_A$
$r^m_{A4} : nonmonotonic\_logic_C \Rightarrow nonmonotonic\_reasoning_A$

$r_{A5}^m : centralized\_architecture_D \Rightarrow \neg distributed\_reasoning_A$
$r_{A6}^m : decentralized\_reasoning_E \Rightarrow distributed\_reasoning_A$

Bill has a local strict rule (r_{B1}^l), which states that 'AmI journal' is about ambient intelligence. Christine's local rule r_{C1}^l expresses that defeasible logic belongs to the family of non-monotonic logics. Daniel has two local rules; r_{D1}^l, which states that blackboard systems are certain types of shared-memory systems; and r_{D2}^l, which states that shared memory systems imply centralized architectures. Ellen's local rule r_{E1}^l expresses that collective reasoning is a type of decentralized reasoning. Furthermore, Ellen has predefined a mapping rule (r_{E2}^m) with the vocabulary used by her fellow student Francis, which associates *collective_intelligence* (used by Francis) with *collective_reasoning*. Francis has a local strict rule (r_{F1}^l) stating that blackboard systems imply collective intelligence.

$r_{B1}^l : ami\_journal \rightarrow ambient\_computing$

$r_{C1}^l : defeasible\_logic \rightarrow nonmonotonic\_logic$

$r_{D1}^l : blackboard\_system \rightarrow shared\_memory$
$r_{D2}^l : shared\_memory \rightarrow centralized\_architecture$

$r_{E1}^l : collective\_reasoning \rightarrow decentralized\_reasoining$
$r_{E2}^m : collective\_intelligence_F \Rightarrow collective\_reasoning_E$

$r_{F1}^l : blackboard\_system \rightarrow collective\_reasoning$

Given a query issued to Adam'a agent (A) about the recommendation of this article, and assuming that

1. Adam's agent manages to establish a connection through the P2P network with the agents of Bill, Christine, Daniel and Ellen, and Ellen's agent can also connect with Francis' agent.
2. The description of the article is shared between the agents of the six fellow students.
3. The trust level order defined by Adam is $T_A = [B, C, E, D]$, where B stands for Bill's agent, C for Christine's agent, D for Daniel's agent, E for Ellen's agent and F for Francis' agent.

the algorithm proceeds as follows: Using the local context theory of A, it is not able to determine if the article must be recommended. It can only derive a positive value for *contextual_reasoning*, but no local answers for *ubiquitous_computing, nonmonotonic_reasoning* and *distributed_reasoning*. Therefore it proceeds to the next steps, in which it uses the mappings rules of agent A.

Through rule r_{A4}^m, it uses the local knowledge of agent C and computes a positive truth value for *nonmonotonic_reasoning*. For *distributed_reasoning*, agent A has defined two conflicting mapping rules (r_{A6}^m and r_{A5}^m). Through rule r_{A6}^m, the algorithm accesses the local knowledge of E and through E's mapping rule r_{E2}^m, it also uses the knowledge of F to determine that r_{A6}^m is applicable. Trying to evaluate the applicability of r_{A5}^m, which contradicts *distributed_reasoning*, the algorithm uses the local knowledge of D, and finally determines that r_{A5}^m is also applicable. As there is no priority relation between the two conflicting rules (r_{A5}^m and r_{A6}^m), the algorithm uses the trust level order of A to determine which of the two rules is stronger. E precedes D in T_A, so the algorithm computes that r_{A6}^m is stronger than r_{A5}^m, and returns a positive value for *distributed_reasoning*. Using the local knowledge of A, the algorithm also computes a positive answer for *contextual_reasoning*, and determines that r_{A2}^d is applicable.

In the next step, the algorithm checks the only rule that contradicts *recommend*, rule r_{A1}^d. Using rule r_{A3}^m and the local knowledge of B, it determines that r_{A1}^d is also applicable.

Using the priority relation $r_{A2}^d > r_{A1}^d$, the algorithm determines that r_{A2}^d overrides r_{A1}^d, and therefore this article should be recommended to Adam.

6   Conclusion

In this study, we proposed the addition of non-monotonic features in Multi-Context Systems in order to model uncertainty and inconsistency in the distributed context knowledge. The model that we described represents contexts as defeasible local theories in a P2P system, and mappings as defeasible rules, and uses a preference relation to resolve the potential conflicts that arise from the interaction of the distributed theories through the mappings. We, then, described a distributed query evaluation algorithm for MCS, and analyzed its formal properties with respect to termination, complexity and the possibility to create an equivalent global defeasible theory from the distributed contexts. Finally, we described the use of the algorithm in a use case scenario from the Semantic Web domain. Part of our ongoing or future work includes:

- Implementing the algorithm in Logic Programming, using the equivalence with Defeasible Logic, and the well-studied translation of defeasible knowledge into logic programs under Well-Founded Semantics [19].
- Studying the equivalence between *non-acyclic* MCS and global defeasible theories constructed in the way described in Section 4, using variants of Defeasible Logic that handle loops in the theory.
- Studying alternative methods for conflict resolution, which differ in the way that a peer evaluates the answers returned by its acquaintance peers; for example we could associate the quality of an answer not only with the trust level of the queried peer, but also with the confidence of the queried peer on the answer it returns (strict answer/defeasible answer).
- Extending the algorithm to support overlapping vocabularies, which will enable different context theories to use elements of common vocabularies (e.g. URIs).
- Studying more applications in the Ambient Intelligence and Semantic Web domains, where the theories may represent ontological context knowledge, policies and regulations.

References

1. McCarthy, J.: Generality in Artificial Intelligence. Communications of the ACM 30(12), 1030–1035 (1987)
2. Buvac, S., Mason, I.A.: Propositional Logic of Context. In: AAAI, pp. 412–419 (1993)
3. McCarthy, J., Buvač, S.: Formalizing Context (Expanded Notes). In: Aliseda, A., van Glabbeek, R., Westerståhl, D. (eds.) Computing Natural Language, pp. 13–50. CSLI Publications, Stanford (1998)
4. Giunchiglia, F., Serafini, L.: Multilanguage hierarchical logics, or: how we can do without modal logics. Artificial Intelligence 65(1), 29–70 (1994)
5. Ghidini, C., Giunchiglia, F.: Local Models Semantics, or contextual reasoning=locality+compatibility. Artificial Intelligence 127(2), 221–259 (2001)
6. Roelofsen, F., Serafini, L.: Minimal and Absent Information in Contexts. In: IJCAI, pp. 558–563 (2005)
7. Brewka, G., Roelofsen, F., Serafini, L.: Contextual Default Reasoning. In: IJCAI, pp. 268–273 (2007)
8. Bernstein, P.A., Giunchiglia, F., Kementsietsidis, A., Mylopoulos, J., Serafini, L., Zaihrayeu, I.: Data Management for Peer-to-Peer Computing: A Vision. In: WebDB, pp. 89–94 (2002)
9. Halevy, A.Y., Ives, Z.G., Suciu, D., Tatarinov, I.: Schema Mediation in Peer Data Management Systems. In: ICDE, pp. 505–516 (2003)
10. Calvanese, D., De Giacomo, G., Lenzerini, M., Rosati, R.: Logical Foundations of Peer-To-Peer Data Integration, pp. 241–251. ACM, New York (2004)
11. Franconi, E., Kuper, G.M., Lopatenko, A., Serafini, L.: A Robust Logical and Computational Characterisation of Peer-to-Peer Database Systems. In: DBISP2P, pp. 64–76 (2003)
12. Calvanese, D., De Giacomo, G., Lembo, D., Lenzerini, M., Rosati, R.: Inconsistency Tolerance in P2P Data Integration: an Epistemic Logic Approach. In: Bierman, G., Koch, C. (eds.) DBPL 2005. LNCS, vol. 3774, pp. 90–105. Springer, Heidelberg (2005)
13. Chatalic, P., Nguyen, G.H., Rousset, M.C.: Reasoning with Inconsistencies in Propositional Peer-to-Peer Inference Systems. In: ECAI, pp. 352–356 (2006)
14. Benerecetti, M., Bouquet, P., Ghidini, C.: Contextual reasoning distilled. JE-TAI 12(3), 279–305 (2000)
15. Bikakis, A.: Distributed Reasoning with Conflicts in a Peer-to-Peer Setting (2008), http://www.csd.uoc.gr/~bikakis/P2PDR.pdf
16. Antoniou, G., Billington, D., Governatori, G., Maher, M.J.: Representation results for defeasible logic. ACM Transactions on Computational Logic 2(2), 255–287 (2001)
17. Governatori, G., Maher, M.J., Billington, D., Antoniou, G.: Argumentation Semantics for Defeasible Logics. Journal of Logic and Computation 14(5), 675–702 (2004)
18. Maher, M.J.: A Model-Theoretic Semantics for Defeasible Logic. In: Paraconsistent Computational Logic, pp. 67–80 (2002)
19. Antoniou, G., Billington, D., Governatori, G., Maher, M.J.: Embedding defeasible logic into logic programming. Theory Pract. Log. Program. 6(6), 703–735 (2006)

Personal Agents in the
Rule Responder Architecture

Benjamin Larry Craig and Harold Boley

Faculty of Computer Science, University of New Brunswick
Institute for Information Technology, National Research Council of Canada
Frederiction New Brunswick, Canada
ben.craig@unb.ca
harold.boley@nrc-cnrc.gc.ca

Abstract. Rule Responder is an intelligent rule-based system for collaborative teams and virtual communities that uses RuleML as its knowledge interchange format. This multi-agent infrastructure allows these virtual organizations to collaborate in an automated manner. It is implemented as a Web Service application on top of Mule, an Enterprise Service Bus. It supports rule execution environments (rule/inference engines) such as Prova and OO jDREW. Rule Responder implements an effective methodology and an efficient infrastructure to interchange and reuse knowledge bases (ontologies and rules). The paper describes the design decisions for the personal agent architecture of Rule Responder. A comparison between our distributed rule bases and a centralized rule base is given. An online use case for Rule Responder, applied to the organization of a symposium, is demonstrated.

1 Introduction

Person-centered and organization-centered profile descriptions using Web 3.0 (Semantic Web plus Web 2.0) techniques are becoming increasingly popular. They are often based on the Resource Description Framework (RDF) and include Friend of a Friend (FOAF)[1], Semantically-Interlinked Online Communities (SIOC)[2], and the ExpertFinder Initiative[3]. Recent work on FindXpRT [BLB⁺] generalized fact-based to rule-based profiles to capture conditional person-centered metadata such as the right phone number to call a person depending on the time, the topic, the caller, or the urgency.

Rule Responder [PBKC07, BP07, Cra07] is a service-oriented middleware tool that can be used by virtual organizations for automated rule-based collaboration [PKB07]. Distributed users (humans or agents) can interact with Rule Responder by query-answer conversations or negotiation and coordination protocols. Rule Responder agents will process events, queries, and requests according to their rule-based decision and behavioral logic. It can also delegate subtasks to

---

[1] http://www.foaf-project.org/
[2] http://www.w3.org/Submission/sioc-spec/
[3] http://wiki.foaf-project.org/ExpertFinder

N. Bassiliades, G. Governatori, and A. Paschke (Eds.): RuleML 2008, LNCS 5321, pp. 150–165, 2008.

other agents, collect partial answers, and send the completed answer(s) back to the requester. The communication protocol used between the architectural components of Rule Responder (e.g., external, personal, and organizational agents) is Reaction RuleML [PKB07]. The Rule Responder Technical Group[PBKC] of RuleML is focused on implementing use cases that require the interchange of rule sets and a querying service. The use cases demonstrate rule-based collaboration in a virtual organization. A virtual organization consists of a community of independent and often distributed (sub)organizations, teams or individual agents that are members of the virtual organization. Typical examples are virtual enterprises, virtual (business) taskforces, working groups, project teams, or resource-sharing collaborations as in, e.g., grid computing or service-oriented computing (SOC).

One specific use case that will be explained in detail throughout the paper is the organization of the RuleML-2008 Symposium. Besides the contribution of an intelligent autonomous agent layer on top of the current Semantic Web, the use case demonstrates rule interchange between rule inference services using a common rule exchange format (RuleML/Reaction RuleML). It also implements a scalable and flexible communication protocol, a service-oriented architecture, and an object broker middleware solution based on enterprise service technologies.

The rest of the paper is structured as follows. Section 2 explains how Rule Responder is implemented as a rule-based multi-agent infrastructure. Section 3 details the real world use case of Rule Responder for organizing a symposium. Section 4 discusses the organizational agent architecture of Rule Responder. Section 5 describes the personal agent architecture. Section 6 compares the differences between distributed rule systems and a centralized rule system. Section 7 concludes the paper. Appendix A contains rule sets for two of our implemented personal agents described in the use case.

2   Rule Responder as a Rule-Based Multi-agent Infrastructure

Rule Responder's architecture realizes a system of personal agents (PAs) and organizational agents (OAs), accessed by external agents (EAs), on top of an Enterprise Service Bus (ESB) communication middleware. The semi-autonomous PAs and OAs are implemented by (an instance of) a rule engine each, which acts as the inference and execution environment for the rule-based decision and behavioral logic of that semi-autonomous agent. The rule-based PAs represent, as their 'dynamic profiles', all of the participating human members of the virtual organization modeled by Rule Responder. An OA constitutes an intelligent filtering and dispatching system, using a rule engine execution environment for either blocking incoming queries or selectively delegating them to other agents. The communication middleware implements an Enterprise Service Bus (ESB) supporting various transmission protocols (e.g., JMS, HTTP, SOAP). The EAs can interact with the Rule Responder-enabled virtual organization via its public communication interface (e.g., an HTTP endpoint interface to an OA as the "single point of entry"). The current development API of Rule Responder uses a Web browser (Web form) for human-machine communication. A new production interface has been created by the Rule Responder Technical Group [PBKC], which will be integrated into the RuleML-2008 Rule Responder webpage. It allows the translation of EA queries from Attempto Controlled English to Reaction RuleML. The ESB implementation for Rule Responder is Mule [BCC+], an efficient open source communication middleware.

Rule Responder blends and tightly combines the ideas of multi-agent systems, distributed rule management systems, as well as service-oriented and event-driven architectures. Rule Responder has the above-discussed three types of agents (PAs, OAs, EAs) which all need to communicate with each other through the Mule ESB. In our current hierarchical use cases, agent-to-agent communication must go through the organizational agent. In particular, when an external agent asks a question to an organization, the external agent does not (need to) know any personal agent such as the one that might ultimately answer the query. Instead, the query must be sent to the organization's OA, which will then delegate it to an appropriate PA. The current Rule Responder use cases do not require direct PA-to-PA communication, but the Rule Responder architecture as described in this paper allows the evolution of our system towards such 'horizontal' (peer-to-peer) communication. The query delegation process and PA-to-PA communication will be described in detail in section 4.2.

While no other rule-based multi-agent ESB system seems to be deployed, Rule Responder could be compared to the Java Agent Development Framework (JADE) [BCR+]. JADE implements an agent-to-agent communication framework, where JADE agents can be distributed over different computers similar to Rule Responder agents. JADE agents can also be made into rule-based agents with the use of the Java-based JESS [FH]. Rule Responder agents can use any rule engine; currently OO jDREW [BC] (Object Oriented java Deductive Reasoning Engine for the Web) and the PROVA [KPS] distributed Semantic Web rule engines are used. Extensions to Rule Responder could allow communication with JADE agents, realized with an XSLT translator from Reaction RuleML to FIFA-ACL. A project similar to JADE is JASON [HBb] which provides a Java-based platform for the development of multi-agent systems. JASON has already created an environment where JASON agents can communicate with JADE agents. [HBa]. We envision that Rule Responder will follow JASON here and interoperate with JADE. This would create interesting communication chains and synergies between several multi-agent infrastructures.

3   RuleML-2008 Use Case Description

One group of use cases created to demonstrate Rule Responder is the organization of meetings such as the RuleML Symposium series, which is an example of a virtual organization that requires online collaboration within a team. Rule Responder started to support the organizing committee of the RuleML-2007 Symposium [Cra07] and was further developed to assist the RuleML-2008

Symposium. The RuleML-2008 use case consists of fully functional knowledge bases for personal agents[4] two of which are partially listed in appendix A. These use cases strive for embodying responsibility assignment, automating first-level contacts for information regarding the symposium, helping the publicity chair (see appendix A.2 for implementation) with sponsoring correspondence, helping the panel chair with, managing panel participants, and the liaison chair (see appendix A.1 for implementation) with coordinating organization partners. They could also aid with other issues associated with the organization of a meeting, including presentation scheduling, room allocation, and special event planning.

The RuleML-2008 use case utilizes a single organizational agent to handle the filtering and delegation of incoming queries. Each committee chair has a personal agent that acts in a rule-governed manner on behalf of the committee member. Each agent manages personal information, such as a FOAF-like profile containing a layer of facts about the committee member as well as FOAF-extending rules. These rules allow the PA to automatically respond to requests concerning the RuleML-2008 Symposium. Task responsibility for the organization is currently managed through a responsibility matrix, which defines the tasks committee members are responsible for. The matrix and the roles assigned within the virtual organization are defined by an OWL (Ontology Web Language) Lite Ontology. The Pellet [HPSM] reasoner is used to infer subclasses and properties from the ontology.

External agents and the RuleML-2008 agents can communicate by sending messages that transport queries, answers, or complete rule sets through the public interface of the OA (e.g., an EA can use an HTTP port to which **post** and **get** requests can be sent from a Web form). The standard protocol for intra-transport of Reaction RuleML messages between Rule Responder agents is JMS. HTTP SOAP is used for communication with external agents, such as Web services or HTTP clients.

4   Organizational Agent Architecture

Organizational agents are used to describe the organization as a whole; for example, an OA contains a knowledge base that describes the organization's policies, regulations, and opportunities. This knowledge base contains condition/action/event rules as well as derivation rules. An example query that the OA can answer for the RuleML-2008 Symposium is: "Who is the contact responsible for the symposium's panel discussion?" When a RuleML-formalized version of this query is received by the OA, this agent must first determine who the correct contact person is for the panel discussion. When the correct contact person for the panel discussion has been selected, the OA delegates the query to that committee member's personal agent. The PA will then respond with the member's name and contact method (e.g., email or telephone number, depending on contact preferences in their FOAF-like profile). Alternatively, if that contact person was on vacation or currently busy, then the PA would respond back to

---

[4] http://www.ruleml.org/RuleML-2008/RuleResponder/index.html

the OA that the contact person is unavailable. If the first-line contact person cannot be reached, then the OA will use the responsibility matrix (i.e., which committee members are responsible for certain tasks, and what members can fill their role if they are unavailable) to try to contact the next PA. This is one way that Rule Responder can act in an automatic process by chaining subqueries that find the best contact person at the time the original query is posed. The responsibility matrix is a method that the OA utilizes for query delegation; an in-depth look at query delegation will be presented in section 5.1. This paper does not focus on the OAs of Rule Responder [PBKC07] but rather on its PAs, which will be discussed next.

5   Personal Agent Architecture

The personal agents used by Rule Responder contain FOAF-extending profiles for each person of the organizational team. Beyond FOAF-like facts, person-centric rules are used. All clauses (facts and rules) are serialized in Naf Hornlog RuleML [HBG+], the RuleML sublanguage for Horn logic (allowing complex terms) enriched by Naf (Negation as failure). These FOAF-extending profiles have access to RDF (BibTeX, vCard, iCard, Dublin Core) and RDFS/OWL (role and responsibility models). The RuleML-2008 Symposium use case [PBC] assists each organization committee member by an implemented personal agent. So the panel chair, general chair, publicity chair, etc. each have their own PA. Each PA contains a knowledge base that represents its chair's responsibilities to answer corresponding queries. For example, the query "What benefits would I receive for sponsoring the symposium with 500 dollars as opposed to 1000 dollars" will be delegated to the publicity chair's agent (see appendix A.2) because it deals with sponsoring for the symposium.

5.1   Query Delegation to Personal Agents

Query delegation is done by the organizational agent, but the personal agents can help the OA in this responsibility. Currently, in the RuleML-2008 use case, task responsibility in the symposium organization is managed through a responsibility matrix, which defines the tasks that committee members are responsible for. The matrix, defined by an OWL Lite Ontology, assigns roles to topics within the virtual organization. As an example, query delegation for sponsoring topics is determined by assigning the publicity chair role to the sponsoring topic in the responsiblity matrix. The ontology also defines which committee chairs can fill in for which other ones. For example, if the publicity chair cannot be reached for queries regarding sponsoring, the general chair can step in and answer such queries. Should there be still no unique PA to delegate a query to, the OA needs to make a heuristic delegation decision and send the query to the PA that most likely would be able to answer the query. For example, if a query about media partners was sent to the OA, it could decide to delegate the query to the publicity chair's PA rather than the liaison chair's PA. Only if the publicity chair's PA was unable to answer the query, would the OA then delegate the query to the liaison chair's PA.

The PAs can help the OAs in query delegation by advertising what kind of queries they can answer via FOAF-like metadata in their knowledge bases. This FOAF-like data could tell the OA what kind of queries the agent is able to solve. This would allow the OA to not have to rely on an assignment matrix defined by an ontology. Such an OA would implement an expert finder approach[BP07] where autonomous agents would search for an expert chair that can answer their queries. A single PA might not be able to answer a query as a whole. Another extension to query delegation would thus be query decomposition, followed by delegation of its decomposed parts to multiple PAs, and finally re-integration of the PAs' answers.

For example, the following rule is currently used to respond to symposium sponsoring queries.

```
sponsor(contact[?Name,?Organization],
        ?Amount:integer,
        results[?Level,?Benefits,?DeadlineResults],
        performative[?Action]) :-
    requestSponsoringLevel(?Amount:integer,?Level),
    requestBenefits(?Level,?Benefits),
    checkDeadline(?DeadlineResults),
    checkAction(?Action,?Level,?Amount:integer).
```

The contact term has two variables, one being the person's name and the other their organization that may want to sponsor the symposium. The amount variable captures how much the organization considers to sponsor. The results term has three variables: for the level of sponsoring (e.g., gold or platinum), the benefits for sponsoring that amount of money, and to find out if the deadline for sponsoring has passed or not. The last term is the performative (action) the publicity chair's agent should take because of the sponsoring query. For this rule to be solved its four premises must be fulfilled.

The first premise will calculate what level the sponsor will receive for donating the amount to the symposium. The second premise will check the benefits that the sponsor will receive. The third premise will determine if the deadline for sponsoring has passed or not. The last premise will conclude what the publicity chair's PA will do in response to the sponsoring query. All of these premises could be solved by the publicity chair's PA. Otherwise, the query would have to be decomposed; for example, in this query the publicity chair's PA may not have the most current date for the deadline for sponsoring or it may not have the newest sponsoring levels or benefits. In order for query decomposition to occur, either the OA has to decompose the query or the PA has to decompose it.

If the OA decomposes the query, it must delegate each premise to a PA. Then the PAs will each solve their premises and send the partial answers back to the OA. The OA will then combine the answers into a single answer. If the PA decomposes the query then the PA would be responsible for finding and

delegating the premises to other PAs. The PA would finally collect all of the
partial answers and send the complete answer back to the OA. Rule Responder
could be extended to return conditional facts (rules) as results of queries. For
example, a rule in response to a sponsoring query could be "If sponsoring takes
place before the August 31st deadline, then benefits will be obtained".

5.2   Performatives

Reaction RuleML is a general, practical, compact and user-friendly XML-seriali-
zed sublanguage of RuleML for the family of reaction rules. It incorporates
various kinds of reaction rules as well as (complex) event/action messages into
the native RuleML syntax using a system of step-wise extensions. For Rule
Responder we use Reaction RuleML as our interchange language between agents.
The following is an example of a communication message that would be delegated
to the publicity chair because the content contains a query for sponsoring the
symposium.

```
<RuleML xmlns="http://www.ruleml.org/0.91/xsd"
 xmlns:xsi="http://www.w3.org/2001/XMLSchema-instance"
 xsi:schemaLocation="http://www.ruleml.org/0.91/xsd
 http://ibis.in.tum.de/research/ReactionRuleML/0.2/rr.xsd"
 xmlns:ruleml2007="http://ibis.in.tum.de/projects/paw#">
   <Message mode="outbound" directive="query-sync">
     <oid><Ind>RuleML-2008</Ind></oid>
     <protocol><Ind>esb</Ind></protocol>
     <sender><Ind>User</Ind></sender>
     <content>
       <Atom>
         <Rel>sponsor</Rel>
         <Expr>
             <Fun>contact</Fun>
             <Ind>Mark</Ind>
             <Ind>JBoss</Ind>
         </Expr>
         <Ind type="integer">500</Ind>
         <Expr>
             <Fun>results</Fun>
             <Var>Level</Var>
             <Var>Benefits</Var>
             <Var>DeadlineResults</Var>
         </Expr>
         <Expr>
             <Fun>performative</Fun>
             <Var>Action</Var>
         </Expr>
       </Atom>
     </content>
   </Message>
</RuleML>
```

The document is encapsulated by `<RuleML>` tags, and the message is contained within `<Message>` tags. The attributes for a message are mode and directive. The mode is the type of a message, either an inbound or an outbound message. The directive is an attribute defining the pragmatic wrapper of the message for example it could be a FIPA-ACL performative. The `<oid>` is the conversation id used to distinguish multiple conversations and conversation states. The `<protocol>` is the transport protocol such as HTTP, JMS, SOAP, Jade, and Enterprise Service Bus (ESB). The `<sender>`/`<receiver>` tags are the sender/receiver of the message. The `<content>` is the message payload transporting a (Reaction) RuleML query, an answer, or a rule base.

The directive attribute corresponds to the pragmatic performative, i.e. the characterization of the message's communication wrapper. External vocabularies defining these performatives could be used by pointing to their conceptual descriptions. The typed logic approach of RuleML enables the integration of external type systems such as Semantic Web ontologies or XML vocabularies. A standard nomenclature of pragmatic performatives is defined by the Knowledge Query Manipulation Language (KQML) and the FIPA Agent Communication Language (ACL). They specify several message exchange/interaction protocols, speech act theory-based performatives, plus content language representations. Other vocabularies such as OWL-QL or the normative concepts of Standard Deontic Logic (SDL) to define, e.g., action obligations or permissions and prohibitions, might be used as well.

5.3   Shared Knowledge between Personal Agents

Shared knowledge is an important aspect of Rule Responder because sharing rules improves flexibility and reduces redundancy. Rules and facts may be relevant to several PAs. The obvious central location for such shared clauses is the OA, who knows where each PA is located. In the RuleML-2008 use case, personal agents can share such OA-centralized knowledge. For example, the following rule can be shared via the OA to obtain contact information for each chair in the symposium's committee.

```
contactChair(?Meeting, ?Chair, ?FirstName, ?LastName,
            ?Title, ?Email, ?Telephone) :-
  person(
  symposiumChair[?Meeting, ?Chair],
  foafname[firstName[?FirstName],lastName[?LastName]],
  foaftitle[title[?Title]],
  foafmbox[email[?Email]],
  exphones[telephoneNumbers[office[?Telephone],cellPhone[?]]]
```

5.4   Query Answering for Personal Agents

In some cases, the OA can try to solve a query from an external agent by itself, but in the following we consider only cases where it delegates queries to

PAs. When a PA receives a query, it is responsible for its answering. If there are multiple solutions to a query, the PA attempts to send an enumeration of as many of the solutions to the OA as possible (it is of course impossible when there are infinitely many solutions). There are different methods for processing multiple solutions to a query. A naive method of the PAs would be to first compute all of the solutions and then send all of the answers back to the OA, one at a time. After the last answer message is sent, an `end-of-transmission` message is sent to let the OA know that there will be no more messages. The main problem with computing all of the answers before sending any of them is obvious: in case of an infinite enumeration of solutions the OA will not receive any answer. The way our implementation addresses the infinite solutions problem is to interleave backtracking with transmission. When a solution is found, the PA immediately sends the answer, and then begins to compute the next solution while the earlier answer is being transferred. When the OA has received enough answers from such a (possibly infinite) enumeration, it can send a `no-more` message to the PA, stopping its computation of further solutions (this is inverse to the `more/ ;` command in sequential Prologs). Once all solutions have been found in a *finite* interleaved enumeration, the PA can send an `end-of-transmission` message.

If a PA is delegated a query and the agent does not have any solutions for it, a `failure` message is sent right away back to the OA. If this situation or a timeout occurs (i.e., the PA is offline and did not respond back to the OA within the preset time period), then the OA can try to delegate the query to another PA to see if *it* is able to solve the query. If no solution can be found in any of these ways, a `failure` message is sent back to the external agent that states that the OA (representing the entire organization) cannot solve the query.

5.5   Communication between Personal Agent and Expert Owner

One problem that can arise when a personal agent works on a query is that the PA may require help or confirmation from its human 'owner'. The PA may not be able to (fully) answer the query until it has communicated with its human owner. The way we approach this problem is to allow PAs to send messages to their owners and vice versa, e.g. in the form of emails. When the owner receives a PA email, he or she can respond to help finding the answer to the external agent. The message format will be in a language the PAs can understand; we propose to again use Reaction RuleML, which can also be generated from Attempto Controlled English [Hir06]. When the personal agent has received the answer from its owner, the PA can use it to complete the answer to the original query.

5.6   Agent Communication Protocols

Rule Responder implements different communication protocols which our agents can utilize. The protocols vary by the number of steps involved in the communication. We try to follow message patterns similar to the Web Service Description

Language (WSDL) [GLS]. For example, there can be **in-only, request-response**, and **request-response-acknowledge** protocols, as well as entire **workflow** protocols. The current Rule Responder use cases primarily focuses on the request-response protocol. The different protocols are explained below with examples.

In-Only: In the **in-only** communication protocol agent$_1$ sends a message to agent$_2$, which then executes the performative. For example, the OA sends a performative to a PA to either assert or retract a clause.

Request-Response: The **request-response** communication protocol starts like **in-only**, but agent$_2$ also sends a response back to agent$_1$. For example, the OA sends a query to a PA, and the PA solves the query and sends the solution back to the OA.

Request-Response-Acknowledge: The **request-response-acknowledge** communication protocol starts like **request-response**, but agent$_1$ then sends an acknowledgment message back to agent$_2$. For example, the OA sends a query to a PA, the PA solves the query and sends the answer back to the OA, and the OA sends an acknowledgement about having received the query's answer to the PA.

Workflows: A **workflow** communication protocol generalizes the above sequential protocols by allowing an arbitrary composition of agent messages from sequential, parallel, split, merge, conditional elements, and loops. For example, the EA sends a query to the OA, the OA decomposes and delegates the subqueries to two PAs, the PAs solve their subqueries sending back the answers to the OA, and the OA integrates the answers and sends the final answer back to the EA.

5.7   Translation between the Interchange Language and Proprietary Languages

Having an interchange language is a key aspect in a distributed rule system. Each agent must be able to understand one common language that every other agent can interpret. The interchange language carries performatives that each agent is able to understand and react to. In Rule Responder we use Reaction RuleML as our interchange language, which carries RuleML queries, answers, and rule bases in the content part of messages. Agents can understand the content of a Reaction RuleML message by interpreting its performative. Since most of our performatives involve rule engines, we need to translate RuleML to proprietary rule engine languages. Each rule engine can have its own syntax and, in order to adopt a rule engine as a Rule Responder agent, there must exist a translator between RuleML and the execution syntax of that rule engine. For example, there exists an XSLT translator from RuleML to Prova, so that Prova can execute RuleML facts and rules. In the case of OO jDREW, there exists a bi-directional translator between RuleML and POSL (RuleML human-readable, Prolog-like syntax) [Bol04].

6   Comparison of a Distributed Rule System vs. a Centralized Rule System

Rule Responder is implemented as a distributed rule system. It connects OAs and PAs so that they can share knowledge and external agents can query this knowledge. Each PA and OA has its own set of rules and facts. The PA's rules (a FOAF-extending profile) correspond to their expert owner while the OA's knowledge describes the virtual organization as a whole. All of the PAs with their rule bases are stored at distributed locations.

In contrast, a centralized rule system would contain all of the facts and rules in one knowledge base or in one centralized location. So, all of the knowledge would be contained in a single file or database. The advantages of a distributed rule system over a centralized system include the ease of distributed maintenance, achieving a fault-tolerant system by using distribution for redundancy, and improved efficiency through distributed processing.

Distributed maintenance allows agents to update their rules and facts without affecting the rule bases of other agents (their consistency, completeness, etc.). If all of the knowledge was stored in one central rule base, problems introduced by one agent would affect the entire system. Distributed maintenance has proved useful when updating the RuleML-2007 use case to the RuleML-2008 one. The PAs are divided into groups for cross-Atlantic maintenance.

Also, if an agent is not currently running (i.e. the server shut down that the agent is running on), the system does not enter a faulty state because not all known agents are running. If a PA is not responding when an OA delegates a query to it, a timeout would be received and either the OA would try another PA that may be able to answer the query or would respond back that the PA is currently offline. When any part of a centralized rule system causes it to go offline, the entire system will be offline until the centralized system starts back online.

In a distributed rule system the knowledge is spread over many different physical locations and communication overhead can become a problem, but the overhead may not be noticeable to external agents. Rules execute faster when there are less clauses for the engine to process, so our distributed approach improves efficiency because we have multiple rule engines working on smaller knowledge modules instead of one rule engine working on a large knowledge base as in a centralized approach.

7   Conclusion

Rule Responder has been used to implement a number of use cases, including the RuleML-2007/2008 symposium organization and the W3C Health and Life Science (HCLS). The middleware used by Rule Responder allows the simultaneous deployment of these use cases. The ESB provides the communication backbone to synchronously or asynchronously interchange messages between multiple agents. RuleML is a descriptive rule interchange language that so far was able to implement all logical structures that were necessary for Rule Responder. For more information about Rule Responder and use case demos, see [PBKC] (section "Use Cases"). Rule Responder is an open source project and is currently using the open source rule engines Prova [KPS] and OO jDREW [BC]. The Rule Responder Technical Group is currently planning to extend the number of rule engines used by the system. Appendix A contains a condensed version of the personal agent's knowledge bases. For a full listing of the PA implementions readers are referred to the use case home page [PBC].

References

[BC]   Ball, M., Craig, B.: Object Oriented java Deductive Reasoning Engine for
            the Web, http://www.jdrew.org/oojdrew/
[BCC+]   Borg, A., Carlson, T., Cassar, A., Cookeand, A., Fenech, S., More.: Mule,
            http://mule.codehaus.org/display/MULE/Home
[BCR+]   Bellifemine, F., Caire, G., Rimassa, G., Poggi, A., Trucco, T., Cortese, E.,
            Quarta, F., Vitaglione, G., Lhuillier, N., Picault, J.: Java Agent DEvelop-
            ment Framework, http://jade.tilab.com/
[BLB+]   Boley, H., Li, J., Bhavsar, V.C., Hirtle, D., Mei, J.: FindXpRT: Find an eX-
            pert via Rules and Taxonomies, http://www.ruleml.org/usecases/foaf/
            findxprt
[Bol04]   Boley, H.: POSL: An Integrated Positional-Slotted Language for Seman-
            tic Web Knowledge (May 2004), http://www.ruleml.org/submission/
            ruleml-shortation.html
[BP07]   Boley, H., Paschke, A.: Expert querying and redirection with rule respon-
            der. In: Zhdanova, A.V., Nixon, L.J.B., Mochol, M., Breslin, J.G. (eds.)
            FEWS. CEUR Workshop Proceedings, vol. 290, pp. 9–22. CEUR-WS.org
            (2007)
[Cra07]   Craig, B.: The OO jDREW Engine of Rule Responder: Naf Hornlog
            RuleML Query Answering. In: Paschke, A., Biletskiy, Y. (eds.) RuleML
            2007. LNCS, vol. 4824. Springer, Heidelberg (2007)
[FH]   Friedman-Hill, E.: JESS - The Rule Engine for the Java Platform, http://
            herzberg.ca.sandia.gov/
[GLS]   Gudgin, M., Lewis, A., Schlimmer, J.: Web Services Description Language
            (WSDL) Version 1.2 Part 2: Message Patterns, http://www.w3.org/TR/
            2003/WD-wsdl12-patterns-20030611/
[HBa]   Hübner, J.F., Bordini, R.H.: Interoperation between Jason and JADE
            Multi-Agent Systems, http://jason.sourceforge.net/mini-tutorial/
            jason-jade/
[HBb]   Hübner, J.F., Bordini, R.H.: Jason - A Java-based interpreter for an ex-
            tended version of AgentSpeak, jason.sourceforge.net/
[HBG+]   Hirtle, D., Boley, H., Grosof, B., Kifer, M., Sintek, M., Tabet, S., Wagner,
            G.: Naf Hornlog XSD, http://www.ruleml.org/0.91/xsd/nafhornlog.
            xsd
[Hir06]   Hirtle, D.: TRANSLATOR: A TRANSlator from LAnguage TO Rules. In:
            Canadian Symposium on Text Analysis (CaSTA), Fredericton, Canada,
            pp. 127–139 (October 2006)
[HPSM]   Hendler, J., Parsia, B., Sirin, E., More.: Pellet: The Open Source OWL DL
            Reasoner, http://pellet.owldl.com/

[KPS]   Kozlenkov, A., Paschke, A., Schroeder, M.: Prova: A Language for Rule
           Based Java Scripting, Information Integration, and Agent Programming,
           http://www.prova.ws/
[PBC]   Paschke, A., Boley, H., Craig, B.: RuleML-2008, Use Case (2008), http://
           www.ruleml.org/RuleML-2008/RuleResponder/index.html
[PBKC]   Paschke, A., Boley, H., Kozlenkov, A., Craig, B.: Rule Responder: A
           RuleML-Based Pragmatic Agent Web for Collaborative Teams and Vir-
           tual Organizations, http://www.responder.ruleml.org
[PBKC07]   Paschke, A., Boley, H., Kozlenkov, A., Craig, B.: Rule Responder: RuleML-
           Based Agents for Distributed Collaboration on the Pragmatic Web. In: 2nd
           ACM Pragmatic Web Conference 2007. ACM, New York (2007)
[PKB07]   Paschke, A., Kozlenkov, A., Boley, H.: A Homogenous Reaction Rule Lan-
           guage for Complex Event Processing. In: Proc. 2nd International Work-
           shop on Event Drive Architecture and Event Processing Systems (EDA-PS
           2007), Vienna, Austria (September 2007)

A   Rulebases and Queries for Personal Agents

The below POSL listings shorten the OO jDREW-implemented PAs of [PBC].

A.1   Abbreviated Version of the Liaison Chair Agent

```
%%% FOAF-like facts concerning the Liaison Chair
% One of three Liaison Chairs is shown here
person(
  symposiumChair[RuleML_2008,Liaison],
  foafname[firstName[Mark],lastName[Proctor]],
  foaftitle[title[Dr]],
  foafmbox[email[MarkATemailDOTcom]],
  exphones[TelephoneNumbers[office[4133],cellPhone[5546]]]).

% Facts about partner organizations
partnerOrganization(RuleML_2008, AAAI).
partnerOrganization(RuleML_2008, W3C).
partnerOrganization(RuleML_2008, BPM_Forum).
% ...other partner organizations omitted for space reasons...

% Rules regarding partner organizations
viewOrganizationPartners(?Partner) :-
  partnerOrganization(?Conference, ?Partner).
% ...more rules omitted...

% Facts Regarding Sponsors
sponsor(RuleML_2008, Vulcan_Inc, Gold).
sponsor(RuleML_2008, Model_Systems, Silver).
sponsor(RuleML_2008, STI_Innsbruck, Bronze).
```

```
% ...other sponsors omitted for space reasons...

% Rules Regarding Sponsors
viewSponsors(?Conference, ?Company) :-
  sponsor(?Conference, ?Company, ?SponsorLevel)
% ...more rules omitted...
```

A.2   Abbreviated Version of the Publicity Chair Agent

```
%%% FOAF-like facts concerning the Publicity Chair
% One of two Publicity Chairs is shown here
person(
  symposiumChair[RuleML_2008,Publicity],
  foafname[firstName[Tracy],lastName[Bost]],
  foaftitle[title[Dr]],
  foafmbox[email[TracyATemailDOTcom]],
  exphones[TelephoneNumbers[office[0314],cellPhone[1234]]]).

% Main sponsor Rule
sponsor(contact[?Name,?Organization],?Amount:integer,
               results[?Level,?Benefits,?DeadlineResults],
               performative[?Action]) :-
  requestSponsoringLevel(?Amount:integer,?Level),
  requestBenefits(?Level,?Benefits),
  checkDeadline(?DeadlineResults),
  checkAction(?Action,?Level,?Amount:integer).

% Rule to check chair's action
checkAction(?Action,?Level,?Amount:integer) :-
  ...see online version...

% Rule to check if deadline has passed
checkDeadline(passed[deadline]):-
  date(?X:integer),
  deadline(sponsoring, ?D:integer),
  greaterThan(?X:integer,?D:integer).

% Rule to check if deadline is ongoing
checkDeadline(onGoing[deadline]):-
  date(?X:integer),
  deadline(sponsoring, ?D:integer),
  lessThan(?X:integer,?D:integer).

% Deadline date
deadline(sponsoring,20080830:integer).
```

```
% Rules to determine sponsor level
requestSponsoringLevel(?Amount:integer,?Level) :-
  sponsoringLevel(rank0,?Level,
  under[us$[?UnderBronzeAmount:integer]]),
  lessThan(?Amount:integer,?UnderBronzeAmount:integer).
requestSponsoringLevel(?Amount:integer,?Level) :-
  sponsoringLevel(rank1,?Level,us$[?BronzeAmount:integer]),
  greaterThanOrEqual(?Amount:integer,?BronzeAmount:integer),
  sponsoringLevel(rank2,silver,us$[?SilverAmount:integer]),
  lessThan(?Amount:integer,?SilverAmount:integer).

% ...rules omitted for each sponsor level...
requestSponsoringLevel(?Amount:integer,?Level) :-
  sponsoringLevel(rank5, ?Level,us$[?EmeraldAmount:integer]),
  greaterThanOrEqual(?Amount:integer,?EmeraldAmount:integer).

% Facts that determine amount of money for each level
sponsoringLevel(rank0, preSponsor,under[us$[500:integer]]).
sponsoringLevel(rank1, bronze,us$[500:integer]).
sponsoringLevel(rank2, silver, us$[1000:integer]).
sponsoringLevel(rank3, gold, us$[3000:integer]).
sponsoringLevel(rank4, platinum, us$[5000:integer]).
sponsoringLevel(rank5, emerald, us$[7500:integer]).

% Rule to request benefits
requestBenefits(?Level,?Benefits) :-
  benefits(?Level,?Benefits).
benefits(preSponsor,benefits[none]).
benefits(bronze, benefits[
                 logo[on[site]],
                 acknowledgement[in[proceedings]]]).
% ...facts omitted for each sponsor levels...
benefits(emerald, benefits[
                   logo[on[site]],
                   acknowledgement[in[proceedings]],
                   option[sponsor[student]],
                   free[registration,amount[3]],
                   logo[in[proceedings]],
                   option[demo],
                   name[all[advance[publicity]]],
                   distribution[brochures[all[participants]]]]).
```

A.3   Selection of Queries for the Personal Agents

The answers are generated online through the EA http interfaces at [PBC].

```
%%% Queries for both chairs
% Query to contact the agents
contactChair(?Conference,?Chair,?FirstName,
                       ?LastName,?Title,?Email,?Telephone)

%%% Queries for Publicity Chair
% Sponsoring query
sponsor(contact[ben,nrc],500:Integer,
   results[?Level,?Benefits,?DeadlineResults],
   performative[?Action])

%%% Queries for Liaison Chair
% Query to view organization partners regardless of event
viewOrganizationPartners(?Partner)

% Query to view sponsors regardless of sponsoring level
viewSponsors(?Conference, ?Company)
```

Semi-automatic Composition of Geospatial Web Services Using JBoss Rules

Raluca Zaharia[1], Laurenţiu Vasiliu[1], and Costin Bădică[2]

[1] Digital Enterprise Research Institute, National University of Ireland, Galway, Ireland
{raluca.zaharia,laurentiu.vasiliu}@deri.org
[2] University of Craiova, Software Engineering Department, Romania
badica_costin@software.ucv.ro

Abstract. The main research focus toward the Semantic Web vision has been so far on conceptualization, while little concern was spent on the efficiency aspect. In this paper we present practical experiences gained with the implementation of a use case from the geospatial domain, in particular a semi-automatic semantic composition of the functionality provided by 7 geospatial web services. Although we obtained the desired results with our initial approach, the performance aspect was unsatisfactory. For this reason we started the investigation into how significant performance improvements could be obtained. We identified rule engines as a solution that would easily fit into our framework and also were mature enough in terms of complexity and execution speed. We provide a new execution solution based on JBoss Rules and compare it with previous implementations.

1 Introduction

For many years companies looked for a way to create enterprise information systems that would support business processes. The emerging approach was Service Oriented Architectures (SOAs), based on the principle of structuring traditional large applications as a collection of services. The goal is to provide dynamics and adaptability to business processes by offering discoverable, reusable and composable services. However, SOA faces a number of challenges that stand in the way of its adoption, such as lack of services metadata and interoperability problems, making the resulting systems difficult to scale. The solution [7] is to extend SOA by adding a semantic layer on top of the existing service stack, resulting in systems more adaptable to change and achieving partial or even total automation through logical reasoning over semantic descriptions of service discovery, mediation, composition and invocation - tasks that become extremely difficult to perform through manual configuration in large systems.

The research reported in this paper has been motivated by a SWING project use case. According to [1], a large number of non-semantic web services are available today within the geospatial domain, but the scarcity of semantic annotation and the lack of a supportive environment for discovery and retrieval make it difficult to employ such services to solve a specific task in geospatial decision making. The aim of the project is to deploy Semantic Web Service (SWS) technology in the geospatial domain.

The geospatial case study is concerned with the management of mineral resources needed as building material for construction sites in France [2]. The solution is a composition of 7 geospatial web services that creates a production-consumption map for

N. Bassiliades, G. Governatori, and A. Paschke (Eds.): RuleML 2008, LNCS 5321, pp. 166–173, 2008.

a geographical area selected by the user. Geospatial web services provide data access and data processing functionalities for geospatial entities. The ratio between aggregate consumption and production on a department scale is an important piece of information for decisions in quarry management. The *production* represents the amount of extracted materials from a geospatial area. A service calculates the production of a department based on the capabilities of the existing quarries in that administrative region. *Quarries* are natural resources extraction sites and a data access service provides this information for a given department. *Consumption* is an average amount of necessary construction materials. A service computes it by multiplying the *population* with a *constant* for average consumption per capita. The selection that a user makes on the map is provided as a pair of (lat, long) coordinates to a data access service that will return the *departments* included in the selected area. The composition determines the departments covered by the area and the average consumption, then the remaining web services are invoked for each department. The total number of invocations is therefore $2 + departments * 5$. The execution of the composition is automatic and the web services that will be actually invoked are only known at runtime, but because there are rules specified by a developer, we consider this a semi-automatic approach to web service composition. The scenario is presented in Figure 1, while the data sources are available at `http://swing.brgm.fr/`.

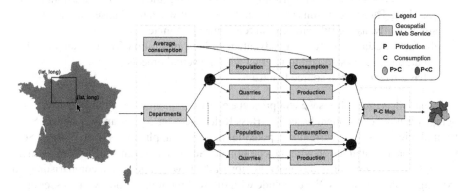

Fig. 1. Use case scenario

Our semantic execution framework is WSMX [11], with abstract state machines (ASM) [3] as the execution formalism. The previous scenarios tested in WSMX focused more on aspects like discovery and mediation for SWS, while the behavioural descriptions (the choreographies) remained quite simple. After implementing the necessary extensions for our composition, the actual execution turned out to be extremely slow. The ASM execution requires many calls to the reasoning component of WSMX in order to ensure correctness, and so query answering with the reasoner proved to be the performance bottleneck. The reasoning component was based on KAON2 [17]. After several optimizations like caching of query results and changing the service description to make it more efficient, the execution speed was greatly improved, but still on the order of minutes for a small geographical surface and below expectations. As an example, for a surface of the size anticipated by the experts who designed the scenario, the results would be returned after approximatively an hour. Our challenge here is to improve the SWS execution performance in WSMX.

de Brujin *et al* [14] survey a set of applications that could be used as reasoners with ontologies expressed in variants of WSML (the language of WSMX). However, [14] discusses only their applicability from the functionality point of view, while reasoning speed is not considered. Also, in [13] the authors mention that most of the tools available for ASMs are not really meant for execution purposes, as the primary role of ASMs is to allow formal specification, simulation and verification of systems. In the related work section, they mention that rule-based systems share a lot of conceptual ideas with ASM implementation, but a translation would require a lot of effort, because the language used by these systems is very different than WSML.

KAON2 is available only as a precompiled binary distribution, thus making impossible to modify or extend it. Since the WSML-Flight variant we used for expressing ontologies and behavioural descriptions is grounded in logic programming (*i.e.* quite close to the rules approach), we decided that it was worth investigating existing rule engines as a solution to our challenge. We chose JBoss Rules for the following reasons:

– first and mainly because of the speed and scalability of the application. JBoss Rules is based on very efficient algorithms[4] for matching rule patterns to the domain object data. These algorithms are especially efficient when the datasets do not change significantly, as the rule engine can remember past matches.
– the facts are Java objects — an advantage to Java integration. The rule engine relies on the JavaBeans standard to interact with the actual objects.
– it has a friendly open source license.

In the following sections we will present a new implementation of the WSMX Choreography Engine component based on JBoss Rules. Section 2 gives an overview of SWS technologies as well as current and potential reasoning applications over WSML. Section 3 covers the interaction with the component. Section 4 presents the new implementation, while section 5 provides an evaluation of the new component in comparison with existing implementations. We conclude with future directions for our work in section 6.

2   Background and Related Work

Vitvar *et al.* [7] note there is scepticism today that semantic languages are too complex and the integration techniques that depend on logical reasoning are a burden for service processing and hinder high performance. They argue that indeed the "logical reasoning can efficiently help resolve inconsistencies in service descriptions as well as maintain interoperability when these descriptions change" and also that "semantics that promote the automation is the key to such integration's flexibility and reliability". They propose a semantic extension to SOA called SESA (semantically enabled SOA) with the aim to fulfil users' goals through logical reasoning over semantic descriptions.

The conceptual semantic service model for SESA is the Web Service Modeling Ontology (WSMO) [9]. Other initiatives such as OWL-S [5] and SAWSDL [6] exist, but WSMO was chosen as it fits best with the design principles [8] of SESA. WSMO describes all relevant aspects of web services through four top level elements: *ontologies* (providing the domain terminology used by other elements), *web services* (describing functional, non-functional and behavioural aspects of services), *goals* (representing user desires, fulfilled by executing a web service) and *mediators* (resolving incompatibilities between the other elements). WSMO is being developed as a complete framework: Web Service Modeling Language (WSML) [10] is a family of ontology languages, Web Service Execution Environment (WSMX) [15] is a reference implementation of the core of the SESA architecture, providing the main intelligence for the integration and interoperation of web services [8] and Web Service Modeling Toolkit (WSMT) [16] is a GUI workbench allowing users to define WSMO elements and to easily interact with the execution environment. Another implementation based on WSMO is IRS-III [18], facilitating interoperability with WSMX through a common system-level API.

The following tools are either already in use to perform reasoning over WSML, or being considered as potential reasoning applications in order to improve the performance of the use case in particular and WSMX in general. KAON2 [17] is an infrastructure capable of providing reasoning over a large subset of OWL-DL and F-logic. KAON2 has been extended with additional meta-level rules in order to handle meta-modeling and it has been used as the main reasoning tool in WSMX. IRIS (Integrated Rule Inference System) [19] is a very recent reasoning engine for Datalog extended with function symbols, unsafe rules, negation, locally stratified or non-stratified programs, XML schema data types and a comprehensive and extensible set of built-in predicates.

JBoss Rules, also known as Drools, is a forward chaining inference-based rule engine using ReteOO, an enhanced and optimised implementation of the Rete algorithm [4] for Object Oriented systems. Another powerful rule engine is Jess [20]. It shares some common features with JBoss Rules, like the fact that it can directly manipulate and reason about Java objects and it is based also on the Rete algorithm. The native language of JBoss Rules, DRL (Drools Rule Language), is simple and lightweight. It supports pluggable dialects allowing other languages to compile and execute expressions and blocks. For this reason we selected Drools as it was easier to adapt our composition to it. Also, Jess is not open source, but it is freely available for academic use.

3   Composition Execution

A choreography is part of the description of a WSMO goal or web service, presenting the communication behaviour *i.e.* how the client interacts with a potential service or how the service communicates with the client, in order to consume the functionality provided by the service [12]. The Choreography Engine (CE) is a WSMX component that handles the execution of a choreography.

One of the main reasons for choosing an ontologized ASM as the process model of CE is the separation of rules and workflows, first because they are better suited for different purposes, but also because separation provides increased reusability and control over both. In general workflows perform some work and here they are represented by specific services (as different workflows can perform a task, there can be many services that provide a certain functionality). Rules on the other hand produce knowledge and so the communication behaviour is defined as a set of interaction steps described by ASM transition rules. This separation is also in accordance with the purpose of the WSMX platform: to provide a degree of automation and to easily solve interoperability issues in the SOA domain; therefore whenever some specific functionality (eg. some arithmetical computations) is required, the composition should include an external service to obtain the result. Thus the behavioural description consists of interaction steps and ASMs are sufficiently expressive, simple and formal to model such communication behaviours.

By ontologized ASM we understand that the state of the ASM execution is represented by an ontology (further referred to as the state ontology). A choreography contains a state signature and transition rules. The state signature defines a set of modes (non-functional properties) for concepts and relations, stating how their instances can be modified. This mechanism is used to ground the mode to an actual service, such as a WSDL endpoint. The execution can later determine if a service should be invoked, based on the available instances of the grounded concept or relation. If so, it will automatically create a request for the service from such instances, perform the invocation and add back the response adapted to WSML instances. The transition rules define how the ASM states are changed by adding, removing and updating instances of the state ontology. They are evaluated in parallel and the updates are executed in parallel as well.

Achieving a goal in WSMX consists of several stages: discovering a matching service for the goal, mediating between ontologies and so forth. The ChoreographyExecution stage, detailed in Figure 2, starts by registering with CE the choreographies of the goal (the requester) and the web service (the provider), as both will be executed. This stage consists in alternatively updating the choreographies states with instances, starting with the goal input data. CE's *updateState* method performs a step in one of the choreographies, depending on the direction: the data is sent from the requester to the provider (R-to-P) or from the provider to the requester (P-to-R). After each update in the R-to-P direction, the returned instances will be used to create service requests and invoke the services. The next phase is calling the *updateState* method with the invocation response instances, but changing the direction. The execution will continue in this manner until one of the following situations occurs: (1) there are no service invocations after an R-to-P update or (2) the provider is in the end state after an P-to-R update. The response ontology for the requester is created with the required instances (if any exist).

4   Contribution

Our new implementation of the Choreography Engine, RulesCE, is based on JBoss Rules. The previous CE implementation uses KAON2 for reasoning (*i.e.* query answering) over the state ontology. Drools can handle both rule execution and query answering, therefore we use it for the reasoning part as well as the execution of the choreography rules. RulesCE ensures a correct translation between WSML and DRL. We preserve the processing model of SWS execution, ASM, and the interaction between the WSMX components is not affected.

Three types of translations are required during the execution phase. (i) The WSML objects (instances, attributes, concepts etc.) are adapted to facts for the rule engine. This is necessary because the WSMO interfaces are not fully compliant with the Java Beans standard required by the facts. (ii) The composition (the rules, the constraints etc.) has

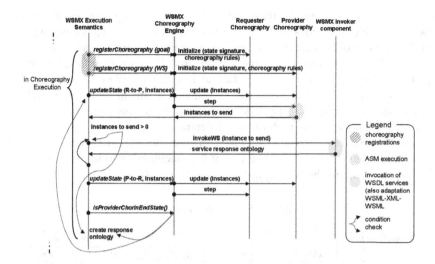

Fig. 2. ChoreographyExecution stage as handled by the overall WSMX execution

to be translated to DRL. WSML allows nested rules, while DRL doesn't. Controlling the order in which JBoss rules should be executed (*e.g.* salience) and extending the conditions of the translated rules overcomes this issue. The choreography rules must be executed in parallel, so the resulted translation must not directly change the facts that correspond to instances. Helper facts store the state ontology changes resulted from the rule activations. A step in the choreography is equivalent to a session run in Drools. (iii) A set of generic queries are added to each rule base. After each step, they extract facts from the rule engine, which are later adapted to WSML instances to be further processed by the CE or other WSMX components.

The rule base will include generic rules that act upon helper facts containing some information about newly created instances. An instance can be described by many helper facts. Because these generic rules have a salience (-1) smaller that any of the translated choreography rules, they will fire afterwards ensuring the helper facts have a unique and correct identifier for the instance they represent.

5 Results and Discussions

Regardless of the tool used to perform the reasoning task, the execution time of a service composition in WSMX is $T = T_1 + T_2 + T_3$. Registering the choreographies, T_1, is a startup overhead which is independent of input. T_2 represents an external phase: creating service requests, waiting for the response, converting back to WSML instances. T_3 denotes the execution time of the choreography rules, a part that is highly influenced by the size of the state ontology. IRIS is now integrated into WSMX so we were able to compare RulesCE with KAON2 and IRIS based implementations.

Table 1 shows experimental results of running the scenario presented in Figure 1 in WSMX. T_1 takes 0.032 s in IRIS and KAON2. The same operation takes longer in

RulesCE because it involves the translation of the choreography rules, but it is constant between executions: 2.282 s. The input data consists of coordinates covering a map area. The amount of data the rules are evaluated against and the overall execution time are highly dependent on the size of the selected area. For a small surface covering one department, the agenda will contain 731 facts at the end of the execution and T_3 will take less than one second. The execution with the different reasoning tools compared here is based on the same principles (the ASM model) and the actions performed by the rules are relatively simple, so the performance difference comes from query answering (or how fast the reasoner is capable to respond). An observation for IRIS: when the number of instances is large, it becomes obvious that the execution hangs in the condition of a rule when determining the binding instances for the variables.

Table 1. Execution performance

Surface size	T_2 (total duration)	Instances in the state ontology	T with reasoning tool:		
			RulesCE	IRIS	KAON2
1department	7 WS (3.48 s)	164	6.58 s	25.434 s	78.578 s
2 departments	12 WS (5.031 s)	321	8.938 s	41.487 s	174.515 s
3 departments	17 WS (6.89 s)	430	10.72 s	55.782 s	314.375 s
14 departments	72 WS (35.655 s)	2058	56.408 s	442.328 s	4231.417 s

6   Conclusions and Future Work

This work has presented a semi-automated composition approach for geospatial web services. As speed can be critical in various applications, choosing the right reasoning tool is vital for the proper execution of any application. We have shown how performance can be improved to execute real business scenarios in reasonable time. A SWING EU project use case scenario has been used for validating the results.

From the research point of view, we have investigated the integration in an unique way of rule engines, SWS and geospatial domain technologies, in order to provide a useful multi-domain application. Several advantages result from the presented approach such as seamless integration, versatility and high decoupling of components. Performing reasoning on top of semantic descriptions with rule engines enhances the usability of the later and increases scalability and performance of the first. The implementation described in this document represents only a prototype, but with some further extensions, the JBoss Rules-based implementation is a fast and reliable reasoning and execution solution in SWS compositions.

Future work will investigate further the trade-off of speed vs. complexity, to implement the existing prototype in a test-bed environment and follow-up possible industrial implementations. One of the main advantages of JBoss Rules over the other reasoners compared in this work is providing a stateful session. With the Datalog-based applications (KAON2, IRIS) state updates cannot be applied directly. Instead, the state ontology is registered (*i.e.* normalized from WSML to Datalog) with the reasoner a large number of times (after every change in the state ontology) in order to ensure correctness of the results. In our first implementation we don't take full advantage of the stateful session as all objects are reasserted at the beginning of each step, but most of the work in JBoss Rules is done when asserting facts, thus we foresee further speed improvements. Another advantage is the possibility of adding general rules (for validation, error reporting, user support etc.) to each rule base. They are defined outside of the composition and are easily reusable, with the rule engine ensuring their fast execution.

Acknowledgments. Science Foundation Ireland grant SFI/02/CE1/I131 and the EU project SWING FP6-26514.

References

1. Semantic Web services INteropeability for Geospatial decision making (SWING), `http://swing-project.org/`
2. Langlois, J., Tertre, F., Bercker, J., Lips, A.: SWING deliverable 1.1 Use Case Definition and IT Requirements (2007), `http://www.swing-project.org/deliverables`
3. Börger, E., Stärk, R.: Abstract State Machines: A Method for High-Level System Design and Analysis. Springer, Heidelberg (2003)
4. Forgy, C.: Rete: A Fast Algorithm for the Many Patterns/Many Objects Match Problem. Artificial Intelligence 19(1), 17–37 (1978)
5. Semantic Markup for Web Services, `http://www.w3.org/Submission/OWL-S/`
6. Semantic Annotations for WSDL and XML Schema, `http://www.w3.org/TR/sawsdl/`
7. Vitvar, T., Zaremba, M., Moran, M., Zaremba, M., Fensel, D.: SESA: Emerging Technology for Service-Centric Environments. IEEE Software magazine 24(6), 56–67 (2007)
8. Vitvar, T., Mocan, A., Kerrigan, M., Zaremba, M., Zaremba, M., Moran, M., Cimpian, E., Haselwanter, T., Fensel, D.: Semantically-enabled Service Oriented Architecture: Concepts, Technology and Application. In: Service Oriented Computing and Applications, vol. 1(2), pp. 129–154. Springer, Heidelberg (2007)
9. Fensel, D., Lausen, H., de Bruijn, J., Stollberg, M., Roman, D., Polleres, A., Domingue, J.: Enabling Semantic Web Services: The Web Service Modeling Ontology, 10.1007/978-3-540-34520-6_5, 57–61
10. Lausen, H., de Bruijn, J., Krummenacher, R., Polleres, A., Predoiu, L., Kifer, M., Fensel, D.: D16.1v0.21 The Web Service Modeling Language, `http://www.wsmo.org/TR/d16/d16.1/v0.21/`
11. Zaremba, M., Moran, M., Haselwanter, T.: D13.4v0.2 WSMX Architecture (2005), `http://www.wsmo.org/TR/d13/d13.4/v0.2/20050613/20050613_d13_4.pdf`
12. Fensel, D., Polleres, A., de Bruijn, J.: D14v1.0. Ontology-based Choreography (2007), `http://www.wsmo.org/TR/d14/v1.0/`
13. Haller, A., Scicluna, J., Haselwanter, T.: D13.9v0.1 WSMX Choreography (2005), `http://www.wsmo.org/TR/d13/d13.9/v0.1/`
14. de Bruijn, J., Feier, C., Keller, U., Lara, R., Polleres, A., Predoiu, L.: D16.2 v0.2 WSML Reasoner Survey, `http://www.wsmo.org/TR/d16/d16.2/v0.2/`
15. WSMX, `http://sourceforge.net/projects/wsmx/`
16. WSMT, `http://sourceforge.net/projects/wsmt/`
17. KAON2, `http://kaon2.semanticweb.org/`
18. Internet Reasoning Service, `http://kmi.open.ac.uk/projects/irs/`
19. IRIS, `http://sourceforge.net/projects/iris-reasoner`
20. Java Expert System Shell, `http://www.jessrules.com/jess/docs/`

A RuleML Study on Integrating Geographical and Health Information

Sheng Gao[1], Darka Mioc[1], Harold Boley[2], Francois Anton[3], and Xiaolun Yi[4]

[1] Department of Geodesy and Geomatics Engineering, University of New Brunswick,
Fredericton, NB, Canada
{sheng.gao,dmioc}@unb.ca
[2] Institute for Information Technology, NRC, Fredericton, NB, Canada
Harold.Boley@nrc.gc.ca
[3] Department of Informatics and Mathematical Modelling, Technical University of Denmark
fa@imm.dtu.dk
[4] Service New Brunswick, Fredericton, NB, Canada
Xiaolun.Yi@snb.ca

Abstract. To facilitate health surveillance, flexible ways to represent, integrate, and deduce health information become increasingly important. In this paper, an ontology is used to support the semantic definition of spatial, temporal and thematic factors of health information. The ontology is realized as an interchangeable RuleML knowledge base, consisting of facts and rules. Rules are also used for integrating geographical and health information. The implemented eHealthGeo system uses the OO jDREW reasoning engine to deduce implicit information such as spatial relationships. The system combines this with spatial operations and supports health information roll-up and visualization. The eHealthGeo study demonstrates a RuleML approach to supporting semantic health information integration and management.

Keywords: Ontologies, RuleML, public health, rules, visualization.

1 Introduction

The Semantic Web improves machine understanding of Web-based information and its effective management. On top of this, semantics-level interoperability among heterogeneous information sources and systems is enabled by the Semantic Web. According to Sheth and Ramakrishnan [1], three kinds of important applications of the Semantic Web are (1) semantic integration, (2) semantic search and contextual browsing, and (3) semantic analytics and knowledge discovery. Nowadays, given the growing number of diseases, improving the performance of health information integration and retrieval becomes very important in analyzing health phenomena (e.g., disease spread). By employing Semantic Web (e.g., Web rule) techniques in health applications, part of the meaning of health information can be captured by machines, thus enabling more precise health information queries, interoperation, etc.

Ontologies, as shared specifications of conceptualizations [2], constitute an important notion in the Semantic Web. Many XML based languages such as RDFS and

N. Bassiliades, G. Governatori, and A. Paschke (Eds.): RuleML 2008, LNCS 5321, pp. 174–181, 2008.
© Springer-Verlag Berlin Heidelberg 2008

OWL have been developed for the representation of ontologies. With the meaning of concepts defined by ontologies, semantic data classification, integration, and deduction can be implemented. For example, the semantic search forms developed by the HealthCyberMap organization link health problems to relevant resources [3] through a medical numerical code or a textual descriptive query. An OWL-encoded ontology was designed to formalize the relationships between similar perinatal concepts in health registries from multiple jurisdictions, enabling the mapping of semantically heterogeneous fields in different databases [4].

However, sometimes it is impossible to explicate all the knowledge as ontologies. For example, spatial relations among disease outbreak locations can be very complex. Implicit spatial information in existing knowledge bases can be deduced by rules. Rules encode machine-interpretable conditional ("if … then …") knowledge for automatic reasoning [5]. Using rule languages can enhance the expressiveness of standard ontology languages [6]. Many different kinds of approaches in combining ontologies and rules have been surveyed (see [7]). Perry et al. [8] discuss the emerging field of extending semantic reasoning from its purely thematic dimension to three dimensions: theme, space, and time.

In health surveillance, efficiently integrating and deducing health information from different sources and flexibly representing it are extremely important. The following issues need to be addressed when integrating health information from different sources and presenting them homogeneously.

Firstly, health information can be collected by hospitals, clinics, government surveys or any other health care facilities in different ways. It is frequently updated with newly reported health cases. Syntactic heterogeneity, (database-)schematic heterogeneity and (terminological-)semantic heterogeneity can exist in these data [9]. These heterogeneities often cause barriers to health information access. Compared with semantic heterogeneity, syntactic heterogeneity and schematic heterogeneity are easier to solve. Bishr et al. [10] even argue that handling syntactic heterogeneity and schematic heterogeneity is not a difficult problem if semantic heterogeneity is solved.

Secondly, health information has the mentioned thematic, spatial and temporal components. Maps are effective tools for the visualization of health information. They give health practitioners an intuitive way of presenting health events. For instance, health practitioners could generate disease distribution maps on administrative areas to understand disease patterns and distributions, and enhance their decision making. Thus, there is a strong need for map-based health information.

The current research mainly addresses the semantic heterogeneity issue of health information and rule knowledge bases. The semantic heterogeneity (different naming methods and interpretations) of health information is focused here rather than its syntactic and schematic heterogeneity. However, since semantic issues are involved in all kinds of heterogeneities, addressing semantic heterogeneity can also help handling these other heterogeneities. An ontology is designed in this study to deal with semantic heterogeneity. Rules are developed for the spatial and thematic retrieval and inference of health information. Moreover, with these rules health information can be rolled up in the spatial and thematic dimensions (e.g., disease categories) to a certain extent from a low level to a high level, and be visualized through maps.

2   Material and Method of the eHealthGeo Study

2.1   Data Description

In this study, the data used contains historical respiratory disease information of the province of New Brunswick (NB), collected by the NB Lung Association. The data include patient incidents (with information such as case id, time, postcode, disease type, age and gender), the spatial boundaries of New Brunswick with respect to province, health region, and census-division levels, as well as the center locations of Canadian postcode areas (using only the first three characters).

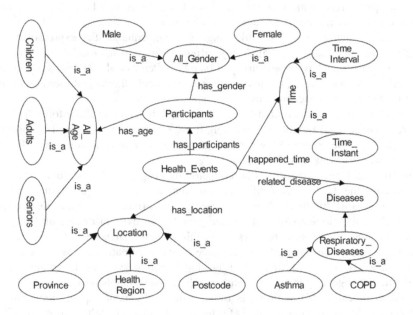

Fig. 1. Portion of the designed ontology

2.2   Ontology Design

The eHealthGeo information is focused on health events, which can describe a variety of cases such as patient incidents, health training services, etc. The following properties associated with health events are considered here: the involved participants' age and gender, the happened time, related disease, and event location. To describe the structure and properties of health events, the relevant vocabulary is designed as an ontology. In particular, subconcepts are connected to superconcepts with "is_a" links. Figure 1 shows a portion of the designed ontology as a directed (property-)labeled graph.

2.3   Knowledge Base Representation

For deductive reasoning, we use the RuleML language to transcribe and refine our ontology as a knowledge base, consisting of facts and rules. RuleML is specified by the Rule Markup Initiative to provide a shared rule markup language, and to support both forward (bottom-up) and backward (top-down) rules in XML for deduction, rewriting, reaction, and further inferential, transformational, and behavioral tasks [11]. As RuleML, which has co-evolved with SWRL [12], SWSL-Rules [13], WRL [14], and RIF [15], is the de facto open language standard for Web rules, using it in our application facilitates knowledge interchange with other tools and applications. The taxonomy ("is_a") backbone of the ontology will be represented as facts, with a rule defining taxonomic inheritance. For example, the central Health_Events class of the ontology in RuleML is transcribed to an event fact with slots for the Participants, Diseases, Location, and Time classes. However, in this RuleML prototype, we directly use the properties of the ontology's Participants class (as the slots age and gender) and omit its happened_time property (explained later).

3   System Architecture and Implementation

3.1   System Architecture

The primary components of the eHealthGeo system include a data tier, a server tier, and a client tier, as shown in Figure 2.

a) Data tier. The heterogeneous geo-referenced health information can be accessed from various databases or files. Following our ontology design, the data can be imported to the server as part of the knowledge base.

b) Server tier. The reasoning engine OO jDREW [16] is responsible for deduction based on facts and rules. Most of the facts are imported from the geo-referenced health information. Further facts are spatial relationships of geospatial objects that are determined through spatial operations. Another important engine on the server side is the map engine GeoTools [17] which generates thematic maps. The map engine joins spatial data with query results from the reasoning engine to create classification maps.

c) Client tier. The client sends user requests to the server and receives query results and thematic maps from the server.

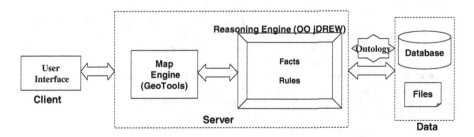

Fig. 2. eHealthGeo system architecture

3.2   Facts in the Knowledge Base

In the following, we use RuleML's POSL presentation syntax [5], which is Prolog-like but also permits 'attribute->value' slots, as in F-logic. While only the (slotted)

Datalog sublanguage of the RuleML family is shown here, the full eHealthGeo study calls for expressivity up to Naf Hornlog [11].

Based on spatial operations and knowledge about the division of administrative boundaries, location facts can be generated. For instance, the place1 with (the region of) the postcode E1V is contained inside place2 within Health_Region_7:

```
inside(place1->E1V;place2->Health_Region_7).
```

Subclass ("is_a") facts are used to represent taxonomic relationships between disease categories. We do not employ a separate (light-weight, subClassOf-only) RDFS or OWL-Lite taxonomy to keep both kinds of knowledge together here, so a rule can be used to define taxonomic inheritance (see below). For example, Chronic Obstructive Pulmonary Disease (COPD) is a kind of respiratory disease, and a respiratory disease is a disease:

```
subclass(disease1->COPD; disease2->Respiratory_Disease).
subclass(disease1->Respiratory_Disease;disease2->Disease).
```

Age facts contain the general notion of the age range of each age group. For example, the age range of adults is defined to extend from age 18 to 64:

```
agerange(agetype->adults;age1->18:Integer;age2->64:Integer).
```

Patient incident facts are generated from geo-referenced health information (database tables or files with column headings (domains) such as id, disease, etc.). For example, we extracted COPD incidents from the NB Lung database:

```
event(id->306947; disease->COPD;  postcode->E1V;
      age->61:Integer; gender->Male).
```

3.3 Rules in the Knowledge Base

Rules are used, e.g., to represent selectors, relational-join queries, and integration views, such as this disease_locator and its subrelations, based on the above facts.

```
disease_locator(id->?id; location->?location; disease->?disease;
                agetype->?agetype; gender->?gender) :-
  event(id->?id; postcode->?postcode; disease->?kindofdisease;
        age->?ageofid:Integer; gender->?gender!?),
  age(agetype->?agetype;agen->?ageofid:Integer),
  inside_closure(place1->?postcode;place2->?location),
  subclass_closure(disease1->?kindofdisease;disease2->?disease).
age(agetype->?agetype;agen->?agex:Integer) :-
  agerange(agetype->?agetype; age1->?age1:Integer;
           age2->?age2:Integer),
  greaterThanOrEqual(?agex:Integer,?age1:Integer),
  lessThanOrEqual (?agex:Integer,?age2:Integer).
inside_closure(place1->?placeA;place2->?placeB) :-
  inside(place1->?placeA;place2->?placeB).
inside_closure(place1->?placeA;place2->?placeC) :-
  inside(place1->?placeA;place2->?placeB),
  inside(place1->?placeB;place2->?placeC).
subclass_closure(disease1->?diseaseA;disease2->?diseaseA).
subclass_closure(disease1->?diseaseA; disease2->?diseaseC) :-
  subclass(disease1->?diseaseA;disease2->?diseaseB),
  subclass(disease1->?diseaseB;disease2->?diseaseC).
```

The disease_locator rule is run in OO jDREW to integrate geographical and health information, permitting optional further requirements by the analyst such as age groups etc. Age rules are used to determine to which age group a certain age belongs. Two partonomy rules define the transitive closure of the inside relation. A non-ground fact and a rule define inheritance via the reflexive-transitive closure of the (disease) subclass relation, representing the taxonomy of disease categories.The knowledge base of the eHealthGeo study is being maintained online at http://gge.athost.net/.

3.4 Health Information Roll-Up and Visualization in the Server

Some implicit spatial relationships can be deduced from the spatial location of geo-referenced health information. To support this 'roll-up' of health information, the spatial relationships between health event locations and administrative boundaries are determined through spatial operations such as point-in-polygon or polygon-in-polygon. For example, these can be employed to determine the health region that a postcode is located in.

Alternatively, running the semantic search query disease_locator(id->?id;location->Health_Region_7;disease->COPD;agetype->adults;gender->Male) against the above knowledge base in OO jDREW produces a number of ?id bindings summed up by the server. Then running the map engine GeoTools generates an intuitive visualization of the health information distribution as a thematic map (cf. Figure 3).

Along with the retrieval and deduction of health information, many other statistical methods, such as crude morbidity ratio and standard morbidity ratio, can also be applied.

4 Results

OO jDREW and GeoTools are used in the eHealthGeo implementation, and both are available as open source Java libraries. OO jDREW is chosen as the deductive reasoning engine because it is the reference implementation of RuleML's Naf Hornlog sublanguage and supports backward and forward reasoning. GeoTools provides standards-compliant methods to manipulate geospatial data [17]. Through the combination of the two engines in our system, health information can be retrieved by combing deductive and spatial processes. Figure 3 shows the classification map of query results for the health region level, COPD, male, and adults. In the spatial and thematic dimensions, the geo-referenced data could be rolled-up and visualized for various administrative boundaries, age, gender and disease categories.

The use of RuleML-based reasoning with OO jDREW enabled us to write a proto-type for integrating geographical and health information. Since we sometimes need to explore huge spatio-temporal data sets, it may become necessary to couple OO jDREW with a database system for storing ground facts. The combination of the reasoning engine (for rules and non-ground facts) with the spatial online analytical processing system (for ground facts) can reduce time latencies in applications. Additional rules, e.g. supporting temporal reasoning via an explicit happened_time slot, will still need to be written in the RuleML / OO jDREW part.

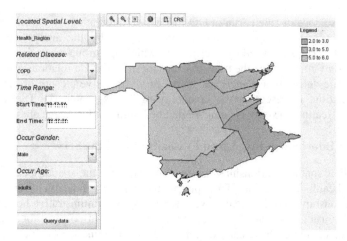

Fig. 3. A thematic map of COPD

5 Conclusion

Recent disease outbreaks have raised growing concern on public health. The World Health Organization, US Centers for Disease Control, and Health Canada are proactively engaged in, e.g., mapping possible viral pandemics for improving global and national health policies. In the present study, we apply ontologies and rules to explore the semantic integration and deduction of geo-referenced health information. Rules are defined to support reasoning on spatial and thematic disease factors, semantic health information integration, and rolling-up health information from a low level to a high level, which is also visualized through thematic maps. This research fosters the use of ontologies and rules in the public health field in order to improve the management of heterogeneous information collected by different health organizations. Our further work will be devoted to the design of a rule-based health decision support system and the improvement on spatial, temporal, and thematic reasoning. The "Health Care and Life Sciences" use case of Rule Responder [18] will be considered as an alternative approach to rule-based integration of heterogeneous health data.

References

1. Sheth, A.P., Ramakrishnan, C.: Semantic (Web) Technology In Action: Ontology Driven Information Systems for Search, Integration and Analysis. IEEE Data Engineering Bulletin 26(4), 40–48 (2003)
2. Gruber, T.R.: A Translation Approach to Portable Ontology Specifications. International Knowledge Acquisition 5(2), 199–220 (1993)
3. HealthCyberMap, Towards a Medical Semantic Web,
   http://www.healthcybermap.org/icd.htm
4. Schuurman, N., Leszczynski, A.: A method to map heterogeneity between near but non-equivalent semantic attributes in multiple health data registries. Health Informatics Journal 14(1), 39–57 (2008)

5. Boley, H.: Are Your Rules Online? Four Web Rule Essentials. In: Paschke, A., Biletskiy, Y. (eds.) RuleML 2007. LNCS, vol. 4824, pp. 7–24. Springer, Heidelberg (2007)
6. Smart, P.D., Abdelmoty, A.I., El-Geresy, B.A., Jones, C.B.: A framework for combining rules and geo-ontologies. In: First International Conference on Web Reasoning and Rule Systems, Innsbruck, Austria (2007)
7. Antoniou, G., Damasio, C.V., Grosof, B., Horrocks, I., Kifer, M., Maluszynski, J., Patel-Schneider, P.F.: Combining Rules and Ontologies: A survey,
   http://rewerse.net/deliverables/m12/i3-d3.pdf
8. Perry, M., Sheth, A., Arpinar, I.B.: Geospatial and Temporal Semantic Analytics. In: Karimi, H.A. (ed.) Encylopedia of Geoinformatics (2007)
9. Bishr, Y.: Overcoming the semantic and other barriers to GIS interoperability. International Journal of Geographical Information Science 12, 299–314 (1998)
10. Bishr, Y.A., Pundt, H., Ruther, C.: Proceeding on the road of semantic interoperability-design of a semantic mapper based on a case study from transportation. In: Proceedings of INTEROPP 1999: 2nd International Conference on Interoperating Geographic Information Systems, Zurich, Switzerland (1999)
11. The Rule Markup Initiative, http://www.ruleml.org/
12. Semantic Web Rule Language (SWRL),
    http://www.w3.org/Submission/SWRL/
13. Semantic Web Services Language (SWSL),
    http://www.w3.org/Submission/SWRL/
14. Web Rule Language (WRL), http://www.w3.org/Submission/WRL/
15. RIF Working Group ,
    http://www.w3.org/2005/rules/wiki/RIF_Working_Group/
16. OO jDREW, http://www.jdrew.org/oojdrew/
17. GeoTools, http://geotools.codehaus.org/
18. Rule Responder, http://responder.ruleml.org/

SBVR Use Cases

Mark H. Linehan

IBM T.J. Watson Research Center
Yorktown Heights, NY 10598
mlinehan@us.ibm.com

Abstract. *Semantics of Business Vocabulary and Rules* (SBVR) is a new standard from the OMG that combines aspects of ontologies and of rule systems. This paper summarizes SBVR, reviews some possible use cases[1] for SBVR, and discusses ways that vocabularies and rules given in SBVR could relate to established ontology standards, to rules technologies, and to other IT implementation technologies. It also describes experience with an SBVR prototype that transforms a subset of SBVR rules types to several types of runtime implementations.

Keywords: SBVR, business rules, rules, semantics, vocabulary, ontology.

1 Introduction

Semantics of Business Vocabulary and Rules (SBVR) [1] is a new standard that formally models business vocabularies and business rules. The "vocabulary" aspect of SBVR defines business-oriented ontologies that include concepts and relationships, definitions, synonyms, closed- versus open-world assertions, and a long list of other features. SBVR rules combine predicate logic with alethic and deontic *modalities* [2]. These capture the "what" of business rules, rather than the "how". They are about modeling business rules, not about executing them.

Accessible descriptions of SBVR are available at [3-8].

What SBVR standardizes is a formal metamodel, and matching XML-based file interchange format, for modeling business vocabularies and rules. The metamodel provides a comprehensive knowledge representation schema that combines ontological concepts with predicate logic, with modal (alethic and deontic) logic, and with features such as objectification of relationships and the ability to tag concepts as either closed- or open-world. The downside of the rich SBVR metamodel is significant complexity and some unresolved integration issues among metamodel aspects.

SBVR explicitly does not formalize a surface representation format for either the vocabulary or the rules, leaving their expression to the invention of tool vendors. The SBVR specification itself employs "Structured English", a form of controlled natural language similar to techniques used by [9-11]. In this paper, SBVR rule examples are given in "Structured English", using several font styles:

> nouns are underlined
> *verbs* are given in italics

---

[1] As used here, the term "use cases" means usage scenarios, not technical use cases as in UML.

N. Bassiliades, G. Governatori, and A. Paschke (Eds.): RuleML 2008, LNCS 5321, pp. 182–196, 2008.
© Springer-Verlag Berlin Heidelberg 2008

<u><u>literal values</u></u> and <u><u>instance names</u></u> are shown with double underlines
keywords are shown in bold font
uninterpreted text is shown in normal font style

Because SBVR is quite different in purpose and benefits from a traditional executable rule system, it seems useful to describe its potential value. This paper summarizes that by describing several SBVR use cases, in two main categories: Modeling Use Cases, and Transformation Use Cases. The paper also describes experiences with a particular top-down transformation technology that uses SBVR.

2   Modeling Use Cases

SBVR uses the term "business" in the Model-Driven Architecture sense of a business-focused, versus implementation-oriented, approach to the vocabulary and rules. One main use of SBVR is to precisely describe concepts and rules in a manner oriented to business users. The prime motivation is clarity in communication among business users and also between business and IT staff.

This section describes three use cases that address this need for clarity by drawing upon the descriptive capability of SBVR. These examples assume that vocabulary and rules are modeled by professionals who have the ability to think abstractly about these ideas, and who are supported by appropriate tools. These professionals may be called "business consultants" or "business analysts" or "business architects" or even "business engineers". Their output is vocabularies and rules expressed in "Structured English" or some other easy-to-understand form. The benefit of using "Structured English" is to enable review and feedback about the vocabulary and rules by ordinary business people.

2.1   Modeling Business Vocabulary and Rules

Consider the contents of legal contracts, which are typically organized as "terms" and "conditions". The terms define and name concepts that are used throughout the contracts. The conditions specify constraints upon the terms (i.e. alethic or structural rules) or upon the behavior of the contract participants (i.e. deontic or behavioral rules). Both the terms and the conditions may be stated more or less precisely. Public laws and regulations typically have a similar structure.

Consider what must happen when a business intends to support a contract, a law, or a regulation. First, a business must understand the legal terms using its own concepts, and deal with any ambiguity. For example, a contract may broadly define a term such as "<u>order</u>". A business might need to consider whether or not a "<u>letter of intent</u>" is an "<u>order</u>". Second, a business must interpret the legal conditions in terms of the business context. For example, a contractual condition might be that "all <u>orders</u> must *be paid within* 30 <u>days</u>". But is that "within 30 days of order receipt" or "within 30 days of order fulfillment" or within 30 days of some other point in the lifecycle of an order? SBVR is about clearly and precisely describing these details, when it matters to a business.

One output of SBVR-style modeling can be a formal business vocabulary, perhaps presented as a dictionary of nouns and relationships. Figure 1 gives an example that shows that "firm order" and "letter of intent" are both kinds of "order". A "letter of intent" is also called an "LOI". A customer can submit either kind of order. In all cases, an order must have at least one item. This kind of business dictionary can help enterprises reduce misunderstandings (e.g. by distinguishing "letter of intent" from "firm order") and clarify expectations (e.g. that an "order" always has at least one "item").

---

customer
Definition: one that purchases a commodity or service
   Source: Merriam Webster Collegiate Dictionary

customer *submits* order *to* company
Definition: **the** customer transmits **the** order to **the** company

firm order
   Definition: order **that** *is final*

letter of intent
Definition: order **that** *is* **not** *final*
Synonym: LOI
Note: A customer may submit an LOI to get a price quote, delivery
      schedule, or other terms of sale.

order
   Definition: customer *request for* items

order *has* item
   Necessity: **Each** order *has* **at least** one item.

Fig. 1. Example Business Vocabulary

Vocabularies may include structural (alethic) rules. They constrain the business concepts themselves, and hence cannot be violated. The necessity rule "**each** order *has* **at least** one item" in Figure 1 is a structural rule.

The other major class of SBVR rules – behavioral or deontic rules – limit or direct business behavior. Figure 2 shows an example, given as both an obligation and a prohibition. The description adds an explanation intended to aid user understanding, but not otherwise interpreted. Behavioral rules can be violated, and can indicate what should happen when they are broken. In Figure 2, the "Enforcement Level" says that that the customer should be notified if an order is shipped when some items are not in inventory.

> A <u>clerk</u> **may** *ship* **an** <u>order</u> **only if all the** <u>items</u> *of* **the** <u>order</u> *are in inventory.*
> Synonymous Statement: **No** <u>clerk</u> **may** *ship* **an** <u>order</u> **if an** *item* **of the** <u>order</u>
>   *is* **not** *in inventory.*
> Description: All parts of the order must be available in the warehouse be-
>   fore any of the order may be shipped.
> Enforcement Level: notify customer

Fig. 2. Example Rules

SBVR enables considerable flexibility in the degree of modeling completeness. For example, vocabulary entries can have informal definitions (e.g. for <u>customer</u>) that use terms not otherwise defined. Noun concepts can also be partially formal: one can say that a noun is a subtype of another noun while stating the distinguishing characteristics informally. Fully formal definitions (e.g. for <u>firm order</u>) enable automated reasoning (e.g. to determine whether a particular <u>order</u> is a <u>firm order</u>).

Despite this flexibility, modeling of any kind is labor-intensive, and hence expensive. Why would this level of detail matter to a business? Large companies, partnerships, and governments have tremendous problems with ambiguity, leading to confusion, inefficiency, loss of time, and sometimes contract or regulatory violations. Another risk is excessive or unpredictable information system development time and effort. The practical impact can be increased business costs, loss of business agility, or even jail for business executives. Those risks often motivate enterprises to make the investment needed to pin down their terms and explicate their business rules.

2.2 Requirements Management

Requirements management is a well-known software engineering discipline involving gathering, articulating, and verifying business and user needs. Requirements typically are documented as a combination of use cases (e.g. as in UML) and text. One problem is that text in any human language is unavoidably ambiguous. Another problem is that most lists of requirements are incomplete.

The need for better requirements management has been identified by many sources. The United States Government Accountability Office (GAO) reported in March 2006 [12], that:

"For example, ill-defined or incomplete requirements have been identified by many experts as a root cause of system failure. As a case in point, we recently reported that the initial deployment of a new Army system intended to improve depot operations was still not meeting user needs, and the Army expected to invest $1 billion to fully deploy the system. One reason that users had not been provided with the intended systems capability was a breakdown in requirements management."

> 1. Training costs are US$300 per student per course.
> 2. All training material will be provided to student on first day of class.

Fig. 3. Example Requirements

Consider the two requirements listed in Figure 3. Some questions that might be asked about these requirements include:

- Is a "course" the same thing as a "class"?
- Who provides the training material?
- Who pays the training costs?

<u>Vocabulary</u>

<u>class</u>
  Synonym: <u>course</u>

<u>class</u> *has* <u>training material</u>
  Synonymous Form: <u>training material</u> *for* <u>class</u>

<u>student</u> *enrolls in* <u>course</u>

<u>student</u> *pays* <u>amount</u> *for* <u>course</u>

<u>Rules</u>

Each <u>student</u> **that** *enrolls in* a <u>course</u> **must** *pay* <u>$300</u> *for* **the** <u>course</u>.

Each <u>school</u> **must** *provide* all <u>training material</u> *for* **a** <u>class</u> *to* **each** <u>student</u> *of* **the** <u>class</u> *on* **the** <u>first day</u> *of* **the** <u>class</u>.

Fig. 4. Requirements Refined as SBVR Vocabulary and Rules

SBVR can help answer such questions by enabling formal definitions of the terms used in the requirements, and of the requirements themselves. Figure 4 shows how an SBVR vocabulary and rules could clarify these requirements. The entry for <u>course</u> defines a noun concept, while the next three entries define fact types (relationships) that are used in the rules. Specifying <u>course</u> as a synonym of <u>class</u> makes explicit that these terms mean the same thing. The first rule identifies who pays for each course. The second rule specifies that the "<u>school</u>" provides the "<u>training material</u>". Detailing the vocabulary and the rules in this manner eliminates many ambiguities.

SBVR-style modeling can thus be seen as an extension of requirements management. When this degree of formality is desired, requirements given informally can be restated as rules that use formal business vocabularies. This should drive out ambiguity, thus making the requirements clearer and more precise. The additional effort needed for formal modeling may be justified in large projects as a way to reduce the risk of project failure as described in the GAO report cited above.

3 Mapping / Transformation Use Cases

Figure 5 illustrates how SBVR fits in the Computation Independent Modeling (CIM) or Business Modeling layer of the OMG's three-layer Model Driven Architecture [14]. That is, SBVR describes business concepts and requirements without addressing their implementation. For example, an SBVR obligation rule specifies <u>what</u> a business must do, possibly under some condition, without saying <u>how</u> the business should do it.

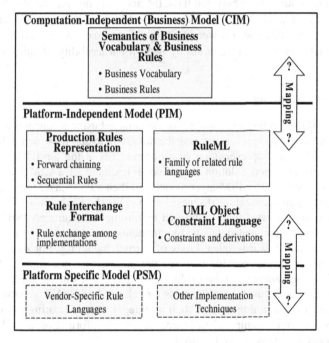

Fig. 5. Rules and Model Driven Architecture

SBVR complements another rules standardization effort at the OMG, called *Production Rules Representation* (PRR) [15]. The latter standardizes a cross-industry model for if-then rules as used in forward chaining and sequential execution. PRR is at the Platform Independent Modeling (PIM) layer because PRR rules define how rules should implement a solution in a vendor-independent manner.

Other activities at the PIM layer include the *RuleML Initiative* [16] and the World Wide Web Consortium's *Rules Interchange Format* (RIF) [17]. Each of these take a vendor-independent approach to some aspect of rules used in solutions. RuleML attempts to define a whole family of rule languages that share core concepts. RIF addresses exchange of rules among different rule systems.

The Object Constraint Language (OCL) [18] of UML also fits at the PIM level. Although not often viewed as a "rule" language, OCL enables modeling of conditional constraints on UML operations, and also derivation of those operations. As discussed below, at least some SBVR rules may be mapped to OCL constraints.

A plethora of rule languages and engines exist both in academia and as commercial products. These fit at the Platform Specific Modeling (PSM) layer, in that there is no ability to exchange rules among implementations. Depending upon how liberally one defines the term "rule", one could extend the view of PSM-layer rules to include things like conditional statements in programming languages, referential integrity constraints in relational database systems, and guards in state machines.

As illustrated in Figure 5, transformations or mapping among these layers is an open issue. Mappings could occur between the CIM and PIM layers, and also between the PIM and PSM layers. The next few sections describe how these transformations may map from an upper layer to a lower layer, or the reverse. "Meet-in-the-middle" use cases also exist. Finally, there is a strong need for traceability of rules across these layers.

3.1 Top-Down Mapping

The term "top-down" refers to a process of defining a business function at a very abstract level, and then successively refining the definition to add more detail. In the MDA framework, an abstract representation of a business fits at the top (CIM or business) layer. When an automated solution is desired, refinement of a CIM business model should lead successively to a PIM-layer model, and then a PSM-layer implementation.

Requirements management, as discussed above, fits the business or CIM layer. Because they are more formal than traditional requirements statements, SBVR vocabulary and rules add rigor and detail to requirements management at this layer. That detail can be a step towards other kinds of formal modeling. For example, SBVR business vocabularies can provide some of the information one finds in UML class models, such as the names and relationships among classes. But vocabularies need not be fully specified to be useful for business modeling. And SBVR vocabulary entries do not address programming details such as integer versus floating point datatypes because these are not meaningful to businesses. Such information must be added at the lower modeling layers, when required.

Top-down transformation of rules often produces a $1:n$ mapping, as illustrated in Figure 6. Business-layer rules map to multiple implementation artifacts. For example, the rule "**Each** order **always** *has* **at least one** item" implies a database table structure with a foreign-key reference among items and orders, and a user interface that supports multiple items per order. The closed-world assertion (via the entry that "'order has item' *is internally closed in* order schema"), implies that the database should have a referential integrity constraint specifying that each order must be referenced by at least one item, and that the user interface should require the entry of at least one item for each new order.

Top-down rules transformation of rules may also generate $n:m$ mappings. These occur when multiple business-layer rules affect a single implementation-layer design feature. For example, both Human Resource rules and national laws may affect security and privacy aspects of personnel systems.

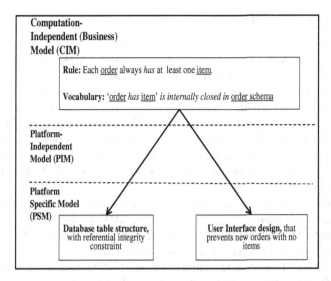

Fig. 6. Example Top-Down Transformation

Most top-down mappings target the automatic creation of executable artifacts, but another valuable output can be test plans. Model-based test generation is an area of active study as exemplified in the A-Most workshop series [19]. SBVR vocabularies and rules can add detail to use cases or other models, thus making automated tests, or test plans, more complete.

In [20], my colleagues and I argue that multiple modeling types, including vocabulary and rules, should be combined to enable comprehensive top-down transformations. For example, consider the rule "A <u>clerk</u> **may** *pack* **an** <u>*order*</u> **only if all the** <u>items</u> *of* **the** <u>order</u> *are on hand*." This rule is perfectly legitimate when taken as a standalone requirement. Coupling the rule with an order handling business process model can enable a more comprehensive transformation. Then the business rule can be understood as a constraint on one or more business process steps. The discussion of the "MDBT Top-Down Transformation", below, describes experiments with this kind of top-down transformation.

Figure 6 avoids the question of what kind of model is appropriate at the PIM layer when doing top-down transformation. The easy answer in the MDA context is some form of UML model. The MDBT toolkit that is described below produces a UML model as an intermediate step. The UML model describes a true platform-independent solution in the sense that it can be mapped to multiple different PSM level implementations. The UML model serves as an indirection layer, isolating the considerations of the business model from the practical details of any particular implementation.

In principle, one could perform top-down transformations directly from a business (CIM) model to a PSM model. Clearly, there is a tradeoff between having a single-step mapping and the two steps shown in figures 6 and 7. Further experience is needed to determine whether the benefits of having the intermediate PIM layer are worth the complexity of a second transformation step.

3.2 Bottom-Up Mapping

Bottom-up mapping of rules is about abstracting rules from software code. This puts it in the space of "Architecture Driven Modernization" (ADM) [21], an OMG task force targeted to analysis and "revitalization" of software.

The concept here is to scan and parse existing software, extract the if-then statements, somehow separate the conditionals that represent business rules from those representing the mechanics of the software, and then recover the original intent of the rules. Clearly this is a significant challenge, even when performed manually. Automating this process, beyond the basics of "mining" the software statements, is very difficult. However, the cost of doing the work manually is so high that even limited automation can have value.

The ADM task force has produced a *Knowledge Discovery Metamodel* (KDM) [22] intended as the basis for such technologies. KDM incorporates a multi-layered structure compatible with MDA ideas, but intended for bottom-up software analysis. The most abstract KDM layer – called the "Abstractions Layer" – explicitly claims alignment with SBVR in order to enable mapping of KDM elements to SBVR rules.

At RuleML 2007, Putrycz and Kark [23] described a tool that partially automates bottom-up mapping of rules. The tool combines ontological techniques, source code scanning, and human input, to reconstruct business-layer rules. The technique holds promise as a way to reduce, but not eliminate, human effort in bottom-up mapping.

The kind of situation shown in Figure 6 makes the bottom-up challenge even greater. In addition to analyzing multiple kinds of software implementations – database table configurations and UI software in this example – the abstraction process needs to notice the common aspects. In this case, the fact that the database expects an item for each order, and the UI requires an item before an order can be entered, should be related to the single common business rule shown in the figure. I am not aware of any automated technologies to accomplish that.

3.3 Meet-in-the-Middle

One of the key questions business leaders ask is how well their operations implement their business policies and rules. Public companies are required to attest to this in formal documents, signed by corporate executives. In the United States, fraudulent attestations can constitute crime and can lead to jail time for the executives. This generates considerable business for corporate auditors.

Imagine a scenario where business rules are formally modeled using SBVR, while business operations are analyzed using bottom-up modeling using something like KDM. Then one could postulate tools and techniques that permit one to compare the expected rules (modeled in SBVR) with the operations (captured in KDM).

This is a fantasy today, since we have fairly limited bottom-up modeling capabilities. If that problem could be solved, then meet-in-the-middle rules comparisons would be very attractive both as a way to avoid executive jail time, and a means to verify software implementations.

3.4 Traceability

Traceability is the maintenance of book-keeping records that link business-layer rules to implementations. These records enable a business to answer these two questions:

1. What is the operational impact of a change to a particular corporate rule?
2. If we change a particular implementation detail, are we still consistent with relevant corporate rules?

Answering these questions is key to regulatory requirement compliance. Automatically maintaining the book-keeping records needed to show traceability has the potential to reduce the risk of non-compliance and the cost of demonstrating compliance.

Traceability may also enable rule explanations (error messages or help text) such as "you can't do that because it would break the corporate rule that ...." The concept is that the "Structured English" used to display rules to rule reviewers can be the basis for providing such explanations to persons who attempt to violate the rules.

4 MDBT Top-Down Transformation

MDBT – Model Driven Business Transformation – is a project at the IBM TJ Watson Research Center to apply automated top-down transformation of business-layer models into executable implementations. MDBT uses "artifact centric" business modeling as an alternative to standard business process modeling. Kumaran describes MDBT in some detail in [24]. In [7] and [8], I describe an extension of MDBT that adds limited SBVR rules modeling. The work is summarized here as an example of top-down business transformation.

4.1 Computation Independent (Business) Model

Figure 7 illustrates the MDBT top-down transformation. Business Analysts define the Business Operation Model (BOM), using an IBM tool called WebSphere Business Modeler [25]. The BOM models artifacts (business information definitions), the artifact lifecycles (defined in a flowchart-like graphical format), the user roles, and associated user tasks. Artifacts are the business documents that record real things or actions. For example, an "order" artifact might represent a customer order.

At the business layer, MDBT models the structure of artifacts and their lifecycle. For example, an order may have attributes such as "order date", and may evolve from an "initialized" state, to subsequent steps such as "complete", "paid", "shipped", "received", "returned", and so forth. One or more user or automated roles are assigned to each lifecycle step. These roles "drive" the evolution of individual artifacts by generating "events" that move artifacts across lifecycle steps. In effect, an artifact lifecycle is a finite state machine that models the states and transitions of the artifact, with the roles identifying what can generate events that trigger the transitions.

The (limited) SBVR prototype takes the artifacts as business vocabulary, and the user roles as noun concepts of a special built-in type "user". In an MDBT extension of SBVR, the events that trigger the transitions form relationships (SBVR fact types) in the model. For example, if a <u>clerk</u> role can *ship* an <u>order</u> in the artifact lifecycle, then the tool internally creates an equivalent relationship "<u>customer</u> *ships* <u>order</u>". These special relationships provide the basis for "restricted permission" rules that extend the Business Operation Model.

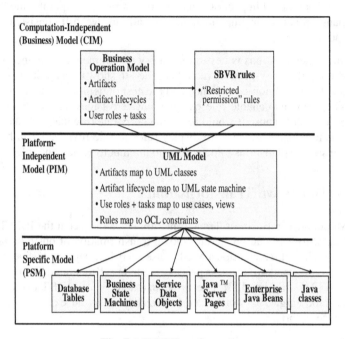

Fig. 7. MDBT Transformation

The rules specify constraints on the actions of the user roles according to conditions that test the attributes of the artifacts. The example given above in Figure 2, "**A** <u>clerk</u> **may** *ship* **an** <u>order</u> **only if all the** <u>items</u> *of* **the** <u>order</u> *are in inventory.*" is a restricted permission rule. It constrains a user (e.g. a UML actor) that fulfills the <u>clerk</u> role from *shipping* an <u>order</u> when the condition is not met. The rule uses a fact type "*<u>clerk</u> ships <u>order</u>*" that semantically links the rule to a "*ship <u>order</u>*" event that may be generated by a "<u>clerk</u>" user role. This application of restricted permission rules is an extension of SBVR concepts to the MDBT artifact lifecycle model.

Each rule refers to an artifact as the second role of the consequent of the rule. For example, rules based on "<u>clerk</u> *ships* <u>order</u>" refer to the <u>order</u> artifact. The rule is evaluated with respect to a particular user who plays the role of <u>clerk</u> and a particular instance of an <u>order</u>. These form the runtime context for the rule.

The condition part of the rule may refer to any vocabulary concepts that are direct or indirect properties of the referenced artifact. Assuming an <u>order</u> has a <u>shipping address</u> and a <u>billing address</u>, then a rule could reference either of these, or properties of them such as <u>zip code</u>. Referenced properties could also include things like <u>prior orders</u> of the same <u>customer</u> if the vocabulary links such properties to the primary <u>order</u> artifact.

4.2   Transformation from Business to Platform Implementation Model

The MDBT transformation converts the combination of the Business Operation Model and the rules to a UML model. The artifacts are converted to UML classes, and the artifact lifecycles become UML state machines. The transitions in these state machines invoke operations on the classes. For example *ship* becomes a UML operation on the order class. The rules become OCL pre-constraints on the corresponding operations. The example rule converts to an OCL constraint that prevents the clerk user role from performing the *ship* operation if the condition is not true.

Figure 8 shows the OCL equivalent of the example rule. Note that the reference to clerk in the original rule is omitted (as here) if the state machine model permits only the clerk role to perform the *ship* action on orders.

```
Context: order:: ship()
pre:      self.items --> forAll (i | i.inInventory)
```

Fig. 8. Example OCL Created from Rules

4.3   Transformation from Platform Implementation Model to Platform Specific Model

The MDBT transformation further converts the UML PIM-layer model to a variety of PSM-layer components as shown in Figure 7. The UML state machines become run-time state machines, using the Business State Machine functions of IBM WebSphere Process Server [26]. The use cases become user interface pages implemented as Java™ Server Pages (JSPs). Transitions in the state machine may be triggered by action buttons in the user interface or by web service requests.

The OCL constraints are implemented as Java methods that test the corresponding conditions. The consequent parts of the rules are implemented in two ways:

1. As guards in the state machines, to prevent the actions from succeeding if the conditions are not met.
2. As user interface functions that enable or disable the corresponding action buttons.

Thus, each business-layer rule automatically constrains both the user interface and the business logic.

4.4   Discussion

These rules are authorization rules in the sense suggested by Goedertier, Mues, and Vanthienen at RuleML 2007 [28]. The "user roles" described here are similar to the "roles" of [28], and the special fact types used here are similar to the "can perform" fact types that represent authorizations in [28]. Goedertier, et. al., develop a schema that permits centralized management of what they call "authorization constraints". A simplified authorization constraint for the example above might be "**It is impossible that a** clerk **ship an** *order* **if there exists an** item *of* **the** order **that** *is* **not** *in inventory*".

Notice that the keyword **impossible** makes this a structural (alethic) rule. I believe behavioral rules – and particularly restrictive permission rules – are a better match to SBVR semantics for authorization constraints.

Experience with this prototype clearly shows that business-layer rules nicely complement and extend the existing MDBT technology. They augment business information and business process modeling as specifications of a business. The rules transformation works smoothly and produces the expected results.

The main problem is the very limited subset of SBVR that is supported so far. Going beyond "restrictive permission" rules will require other transformations. In particular, SBVR structural (alethic) rules map naturally to constraints applied to UML associations rather than to UML operations. For example, a rule such as "Product A should never be cheaper than Product B" (Høydalsvik, G. M. & Sindre [27]) is an alethic rule that constrains the relationship between the concepts <u>Product A</u> and <u>Product B</u>. I hope to address these and other SBVR vocabulary and rule concepts over time.

5   Conclusions

This paper has outlined multiple use cases for business vocabulary and rules as conceived in SBVR. The common theme among these scenarios is the use of SBVR to "raise the abstraction level" of business solutions. Instead of describing rules in terms of implementation concepts such as if-then statements, SBVR captures business requirements – what a business wants. This continues the long-standing Computer Science tradition of searching for more abstract ways to understand technical and business issues.

Figure 5 illustrates the complementary nature of SBVR and implementation-layer rule languages, such as RuleML. Mapping between SBVR and implementation rules exploits the best aspects of each. From a technical viewpoint, defining such mappings can help formalize the SBVR concepts in terms of existing well-defined rule languages, and can enrich rule execution systems with new concepts and new use cases. A "partnership" between SBVR and the traditional rule systems – both from the industrial sector and the academic world -- will benefit all of them.

References

1. Object Modeling Group (OMG): Semantics of Business Vocabulary and Business Rules Specification, Version 1.0 (2007), http://www.omg.org/spec/SBVR/1.0/
2. Halpin, T.: Business Rule Modality. In: EMMSAD 2006, the Eleventh International Workshop on Exploring Modeling Methods in Systems Analysis and Design, Luxembourg (2006), http://www.orm.net/pdf/RuleModality.pdf, http://emmsad06.idi.ntnu.no/EMMSAD06_p3-halpin.pdf
3. Ross, R.: Business Rule Concepts, 2nd edn. Business Rule Solutions, LLC (2005), http://www.brsolutions.com/b_concepts.php, ISBN -941049-06-X
4. Business Rules Community (1997-2007), http://www.brcommunity.com/
5. Chapin, D.: Semantics of Business Vocabulary & Business Rules. In: The W3C Workshop on Rule Languages for Interoperability (2005), http://www.w3.org/2004/12/rules-ws/slides/donaldchapin.pdf, http://www.w3.org/2004/12/rules-ws/paper/85/

6. Chapin, D., Hall, J.: Developing Business Models with SBVR and the BMM. In: The 6th European Business Rules Conference (Dusseldorf 2007) (2007), http://www.eurobizrules.org/Uploads/Files/Chapin_20tutorial.pdf
7. Linehan, M.: Semantics in Model-Driven Business Design. In: SWPW 2006, the 2nd International Semantic Web Policy Workshop, Athens, GA (2006), http://www.l3s.de/~olmedilla/events/2006/SWPW06/programme/paper_02.pdf
8. Linehan, M.: Ontologies and Rules in Business Models. In: VORTE 2007, the 3rd International Workshop on Vocabularies, Ontologies and Rules for The Enterprise, Annapolis, MD (2007)
9. Fuchs, N.E., Schwitter, R.: Attempto Controlled English (ACE). In: CLAW 1996, First International Workshop on Controlled Language Applications. University of Leuven, Belgium (1996), http://www.ifi.unizh.ch/attempto/description/index
10. Bernstein, A., Kaufmann, E.: GINO - A Guided Input Natural Language Editor. In: Cruz, I., Decker, S., Allemang, D., Preist, C., Schwabe, D., Mika, P., Uschold, M., Aroyo, L.M. (eds.) ISWC 2006. LNCS, vol. 4273. Springer, Heidelberg (2006), http://iswc2006.semanticweb.org/items/Bernstein2006tg.pdf
11. Sowa, J.: Common Logic Controlled English, http://www.jfsowa.com/clce/specs.htm
12. GAO, Financial Management Systems – Additional Efforts Needed to Address Key Causes of Modernization Failures. United States Government Accountability Office Report to Congressional Requesters. GAO-06-184 (2006), http://www.gao.gov/cgi-bin/getrpt?GAO-06-184
13. Altwarg, R.: Controlled Languages, an Introduction. Centre for Language Technology website on Controlled Natural Languages, Macquarie University (2006), http://www.shlrc.mq.edu.au/masters/students/raltwarg/clwhatisa.htm
14. Miller, J., Mukerji, J.: MDA Guide Version 1.0.1 (2003); Published by the Object Modeling Group at, http://www.omg.org/docs/omg/03-06-01.pdf
15. Object Modeling Group (OMG), Production Rules Representation, Draft Adopted Specification (2007), http://www.omg.org/cgi-bin/doc?bmi/07-08-01
16. Boley, H., Tabet, S., Wagner, G.: Design Rationale of RuleML: A Markup Language for Semantic Web Rules. In: Proceedings of the Semantic Web Working Symposium (SWWS 2001), Stanford, pp. 381–401 (2001), http://www.ruleml.org/
17. Welty, C., de Sainte Marie, C.: Rule Interchange Format (RIF). World Wide Web Consortium working group report (2006), http://www.w3.org/2005/rules/
18. Warmer, J., Kleppe, A.: The Object Constraint Language: Getting Your Models Ready for MDA, 2nd edn. Addison-Wesley Professional, Boston (2003)
19. A-Most., 4th Workshop on Advances in Model Based Testing (2008), http://kimba.mat.ucm.es/AMOST08/
20. Nayak, N., et al.: Core Business Architecture for a Service-oriented Enterprise. IBM Systems Journal 46(2) (2007), http://www.research.ibm.com/journal/sj/464/nayak.pdf
21. ADM. Architecture Driven Modernization task force at the Object Management Group (OMG), http://adm.omg.org/
22. Object Modeling Group (OMG), Knowledge Discovery Metamodel, Draft Adopted Specification (2007), http://www.omg.org/cgi-bin/doc?ptc/2007-03-15

23. Putrycz, E., Kark, A.: Recovering Business Rules from Legacy Source Code for System Modernization. In: Paschke, A., Biletskiy, Y. (eds.) RuleML 2007. LNCS, vol. 4824. Springer, Heidelberg (2007),
http://2007.ruleml.org/docs/
erik%20putrycz%20-%20RuleML%202007.pdf
24. Kumaran, S.: Model Driven Enterprise. Proceedings of Global Integration Summit, Banff, Canada (2004)
25. IBM WebSphere. IBM WebSphere Business Modeler,
http://www-306.ibm.com/software/integration/wbimodeler
26. IBM WebSphere. IBM WebSphere Process Server,
http://www-306.ibm.com/software/integration/wps/
27. Høydalsvik, G.M., Sindre, G.: On the purpose of object-oriented analysis. ACM SIG-PLAN Notices 28(10), 240–255 (1993),
http://citeseerx.ist.psu.edu/viewdoc/
summary?doi=10.1.1.47.9939
28. Goedertier, S., Mues, C., Vanthienen, J.: Specifying Process-Aware Access Control Rules in SBVR. In: Paschke, A., Biletskiy, Y. (eds.) RuleML 2007. LNCS, vol. 4824, pp. 39–52. Springer, Heidelberg (2007)

Visualization of Proofs in Defeasible Logic

Ioannis Avguleas[1,2], Katerina Gkirtzou[1,2], Sofia Triantafilou[1,2],
Antonis Bikakis[1,2], Grigoris Antoniou[1,2], Efstratios Kontopoulos[3],
and Nick Bassiliades[3]

[1] Computer Science Department, University of Crete, Heraklion, Greece
[2] Institute of Computer Scince (ICS), Foundation of Research and Technology, Hellas
(FORTH), Heraklion, Greece
[3] Department of Informatics, Aristotle University of Thessaloniki,
Thessaloniki, Greece

Abstract. The development of the Semantic Web proceeds in steps,
building each layer on top of the other. Currently, the focus of research
efforts is concentrated on logic and proofs, both of which are essential,
since they will allow systems to infer new knowledge by applying principles on the existing data and explain their actions. Research is shifting
towards the study of non-monotonic systems that are capable of handling
conflicts among rules and reasoning with partial information. As for the
proof layer of the Semantic Web, it can play a vital role in increasing the
reliability of Semantic Web systems, since it will be possible to provide
explanations and/or justifications of the derived answers. This paper
reports on the implementation of a system for visualizing proof explanations on the Semantic Web. The proposed system applies defeasible logic,
a member of the non-monotonic logics family, as the underlying inference system. The proof representation schema is based on a graph-based
methodology for visualizing defeasible logic rule bases.

1 Introduction

The development of the Semantic Web proceeds in steps, building each layer on
top of the other. At this point, the highest layer of the Semantic Web is the
ontology layer, while research starts focusing on the development of the next
layers, the logic and proof layers. The implementation of these two layers is very
critical, since they will allow the systems to infer new knowledge by applying
principles on the existing data, explaining their actions, sources and beliefs.

Recent trends of research focus mainly on the integration of rules and ontologies, which is achieved with Description Logic Programs (DLPs) [1], [2], [3] or
with rule languages like TRIPLE [4] and SWRL [5]. Another interesting research
effort involves the standardization of rules for the Semantic Web, which includes
the RuleML Markup Initiative [6] and the Rule Interchange Format (RIF) W3C
Working Group.

Recently research has been shifted towards the study of non-monotonic systems capable of handling conflicts among rules and reasoning with partial information. Some recently developed non-monotonic rule systems for the Semantic
Web are:

N. Bassiliades, G. Governatori, and A. Paschke (Eds.): RuleML 2008, LNCS 5321, pp. 197–210, 2008.

1. DR-Prolog [7] is a system that implements the entire framework of Defeasible Logic, and is thus able to reason with: monotonic and nonmonotonic rules, preferences among rules, RDF data and RDFS ontologies. It is syntactically compatible with RuleML, and is implemented by transforming information into Prolog.
2. DR-DEVICE [8] is also a defeasible reasoning system for the Semantic Web. It is implemented in CLIPS, and integrates well with RuleML and RDF.
3. SweetJess [9] implements defeasible reasoning through the use of situated courteous logic programs. It is implemented in Jess, and allows for procedural attachments, a feature not supported by any of the aforementioned implementations.
4. dlvhex [10] is based on dl-programs, which realize a transparent integration of rules and ontologies using answer-set semantics.

As for the proof layer of the Semantic Web, it has not yet received enough attention, although it can play a vital role in the eventual acceptance of the Semantic Web on behalf of the end-users. More specifically, for a Semantic Web system to be reliable, explanations and/or justifications of the derived answers must be provided. Since the answer is the result of a reasoning process, the justification can be given as a derivation of the conclusion with the sources of information for the various steps. On the other hand, given a reasoning system is able to provide solid proof explanations, it is important to choose an effective and fully expressive representation of the proof to facilitate agent communication.

In this work we describe a system for visualizing proof explanations on the Semantic Web. The proposed system is based on the implementation presented in [11], [12], which uses defeasible logic [13], a member of the non-monotonic logics family, as the underlying inference system. The proof representation schema adopted by our approach is based on [14], a graph-based methodology for visualizing defeasible logic rule bases.

The rest of the paper is organized as follows: Section 2 presents the basic notions of defeasible logics, focusing on its proof theory. The next section discusses the approach followed for generating and representing visualizations of proofs in defeasible logic, accompanied by an example that better illustrates our methodology. The paper ends with a final section that features concluding remarks and poses directions for future research and improvements.

2   Defeasible Logics

2.1   Basics

Defeasible logics is a simple rule-based approach to reasoning with incomplete and inconsistent information. It is suitable to model situations where there exist rules and exceptions by allowing conflicting rules. A superiority relation is used to resolve contradictions among rules and preserve consistency. Formally, a defeasible theory is a triple $(F, R, >)$ where F is a set of literals, R is a finite set of rules and $>$ is a superiority relation on R. There are three kinds of rules:

- Strict rules denoted $A \rightarrow p$ represent rules in the deductive sense. That is, if the premises of the rule are indisputable, the supported literal holds indisputably as well.
- Defeasible rules denoted $A \Rightarrow p$ represent rules that can be defeated by contradicting evidence. That is, when the premises of the rule hold, the conclusion of the rule holds as well unless there exist stronger conflicting evidence.
- Defeaters denoted $A \rightsquigarrow p$ and are used to defeat some defeasible rules by supporting conflicting evidence.

A superiority relation is an acyclic relation $>$ on R that imposes a partial ordering among elements in R. Given two rules r_1 and r_2, if $r_1 > r_2$, we say that r_1 is superior to r_2 and r_2 is inferior to r_1.

2.2 Proof Theory

A conclusion in D is a tagged literal and may have one of the following form:

- $+\Delta q$, meaning q is definitely provable in D.
- $+\partial q$, meaning q is defeasibly provable in D.
- $-\Delta q$, meaning q has proved to be not definitely provable in D.
- $-\partial q$, meaning q has proved to be not defeasibly provable in D.

In order to prove that a literal is definitely provable, we need to establish a proof for q in the classical sense, that is, a proof consisting of facts and strict rules only, and no other matters need to be taken into consideration.

Whenever a literal is definitely provable, it is also defeasibly provable. In that case the defeasible proof for q coincides with the definite proof. Otherwise, in order to prove q defeasibly in D we must find a strict or defeasible rule supporting q that can be applied. In addition, we must also make sure that the specified proof is not overridden by contradicting evidence. Therefore, we must first make sure that the negation of q is not definitely provable in D. Sequentially, we must consider every rule that is not known to be inapplicable and has head $\sim q$. For each such rule s we require that there is a counterattacking rule t with head q with the following properties:

- t must be applicable at this point.
- t must be stronger than s.

To prove that q is not definitely provable in D, q must not be fact, and every strict rule supporting q must be known to be inapplicable.

If a literal q is proved to be not definitely provable, it is also proved to be not defeasibly provable. Otherwise , in order to prove that a literal is not defeasibly provable we first make sure it is not definitely provable. In addition, one of the following conditions must hold:

- None of the rules with head q can be applied
- $\sim q$ is definitely provable

– There is an applicable rule r with head $\sim q$ such that no possibly applicable rule s with head q is superior to r.

A system attempting to provide a graphical representation of a proof explanation based on defeasible reasoning must incorporate all of the aforementioned cases. The challenge of visualizing such a proof explanation lies in the non-monotonicity of the theory that increases the complexity of a well-established proof explanation in comparison to classic, deductive logics. The system developed is described in more detail in the following section.

3   Method

3.1   Tree–Based Proof Explanation in XML

In order to perform reasoning over a defeasible theory and to provide visualization of respective proofs we used a system proposed in [11], [12]. This approach is based on a translation of a defeasible theory into a logic metaprogram as is defined in [15], [16], that works in conjunction with the logic programming system XSB to support defeasible reasoning. When queried upon a literal, XSB produces a trace of all the successful and unsuccessful paths of the proof explanation. The trace tree is pruned to keep only the necessary information of every proof tree, and the pruned proof explanation is then expressed in XML, a meta-language widely used in the Semantic Web.

In their XML schema, they used a similar syntax to RuleML to represent *Facts* and *Rules*. *Atom* element which refers to an atomic formula is used consisting of two elements, an operator element (Op) and a finite set of Variable (Var) or/and Individual constant elements (Ind), preceded optionally by a not statement (in case representation of a negative literal is required). A *Fact* consists of an Atom that comprises certain knowledge. The last primitive entity of the schema is Rule. In defeasible logic, distinction between two kinds of Rules is provided: *Strict Rules* and *Defeasible Rules*. In the proposed schema every kind of rule is noted with a different element. Both kind of rules consist of two parts, the Head element which constitutes of an Atom element, and the Body element which constitutes of a number of Atom elements.

```
<xsd:element name = "Atom">
  <xsd:complexType>
    <xsd:choice>
      <xsd:sequence>
        <xsd:element name= "Op"/>
        <xsd:sequence minOccurs = "0" maxOccurs = "unbounded">
          <xsd:element name= "Var" minOccurs = "0"/>
          <xsd:element name ="Ind" minOccurs = "0"/>
        </xsd:sequence>
      </xsd:sequence>
      <xsd:sequence>
        <xsd:element name="Not">
          <xsd:complexType>
            <xsd:sequence>
              <xsd:element name= "Op"/>
              <xsd:sequence minOccurs = "0" maxOccurs = "unbounded">
                <xsd:element name= "Var" minOccurs = "0"/>
```

```
                <xsd:element name ="Ind" minOccurs = "0"/>
              </xsd:sequence>
            </xsd:sequence>
          </xsd:complexType>
        </xsd:element>
      </xsd:sequence>
    </xsd:choice>
  </xsd:complexType>
</xsd:element>

<xsd:element name = "Strict_rule">
  <xsd:complexType>
    <xsd:sequence>
      <xsd:element ref= "Head"/>
      <xsd:element ref= "Body"/>
    </xsd:sequence>
    <xsd:attribute name = "Label" type = "xsd:string" use="required"/>
  </xsd:complexType>
</xsd:element>

<xsd:element name = "Defeasible_rule">
  <xsd:complexType>
    <xsd:sequence minOccurs="0">
      <xsd:element ref= "Head"/>
      <xsd:element ref= "Body"/>
    </xsd:sequence>
    <xsd:attribute name = "Label" type ="xsd:string" use="required"/>
  </xsd:complexType>
</xsd:element>

<xsd:element name= "Head">
  <xsd:complexType>
    <xsd:sequence>
      <xsd:element ref= "Atom"/>
    </xsd:sequence>
  </xsd:complexType>
</xsd:element>

<xsd:element name= "Body">
  <xsd:complexType>
    <xsd:sequence>
      <xsd:element ref= "Atom" minOccurs="0" maxOccurs="unbounded"/>
    </xsd:sequence>
  </xsd:complexType>
</xsd:element>

<xsd:element name= "Fact">
  <xsd:complexType>
    <xsd:sequence>
      <xsd:element ref= "Atom"/>
    </xsd:sequence>
  </xsd:complexType>
</xsd:element>
```

Different elements exists also for each type of proof. More specifically, the *Definitely provable* element consists of the *Atom* to be proven and its *Definite proof*, while the *Definite proof* itself consists either of a Strict rule supporting the Atom to be proven with the respective definite proof for each literal in the rule's body. In case the literal in question is a fact the *Definite proof* consists solely of the corresponding *Fact* element. The *Not Definitely provable* element consists of the *Atom* in question and its *Not Definite proof*. The *Not Definite proof* consists of all possible strict rules that support the literal in question and the reason they are blocked (*Blocked* element).

```
<xsd:element name = "Definitely_provable" >
    <xsd:complexType>
      <xsd:sequence>
        <xsd:element ref = "Atom" />
        <xsd:element ref = "Definite_Proof" />
      </xsd:sequence>
    </xsd:complexType>
  </xsd:element>

  <xsd:element name = "Definite_Proof">
    <xsd:complexType>
      <xsd:choice>
        <xsd:sequence>
          <xsd:element ref= "Strict_rule"/>
          <xsd:element ref= "Definitely_provable" minOccurs="0" maxOccurs="unbounded"/>
        </xsd:sequence>
        <xsd:element ref= "Fact"/>
      </xsd:choice>
    </xsd:complexType>
  </xsd:element>

<xsd:element name = "Not_Definitely_provable">
    <xsd:complexType>
      <xsd:sequence>
        <xsd:element ref = "Atom" />
        <xsd:element ref = "Not_Definite_Proof" />
      </xsd:sequence>
    </xsd:complexType>
  </xsd:element>

  <xsd:element name = "Not_Definite_Proof">
    <xsd:complexType>
      <xsd:sequence>
        <xsd:element ref= "Blocked" minOccurs="0" maxOccurs="unbounded"/>
      </xsd:sequence>
    </xsd:complexType>
  </xsd:element>

  <xsd:element name = "Blocked">
    <xsd:complexType>
      <xsd:choice>
        <xsd:sequence>
          <xsd:element ref="Defeasible_rule"/>
            <xsd:choice>
              <xsd:element ref="Superior"/>
              <xsd:element ref="Not_Defeasibly_provable" />
            </xsd:choice>
        </xsd:sequence>
        <xsd:sequence>
          <xsd:element ref="Strict_rule"/>
          <xsd:element ref= "Not_Definitely_provable"/>
        </xsd:sequence>
        <xsd:element name="Not_Superior">
          <xsd:complexType>
            <xsd:sequence>
              <xsd:element ref="Defeasible_rule"/>
            </xsd:sequence>
          </xsd:complexType>
        </xsd:element>
      </xsd:choice>
    </xsd:complexType>
  </xsd:element>
```

A *Defeasibly provable* element consists of the *Atom* to be proven and its *Defeasible proof*. The *Defeasible proof* consists of the applicable rule supporting the *Atom* to be proven and its *Defeasible proof*, followed by a *Not Definitely provable* element concerning the negation of *Atom*, and a sequence of *Blocked* elements for every rule that is not known to be inapplicable and has head the negation of the *Atom* in question.

```
<xsd:element name = "Defeasibly_provable" >
  <xsd:complexType>
    <xsd:choice>
      <xsd:element ref="Definitely_provable"/>
      <xsd:sequence>
        <xsd:element ref = "Atom" />
        <xsd:element ref = "Defeasible_Proof" />
      </xsd:sequence>
    </xsd:choice>
  </xsd:complexType>
</xsd:element>

<xsd:element name = "Defeasible_Proof">
  <xsd:complexType>
    <xsd:sequence>
      <xsd:choice>
        <xsd:element ref= "Strict_rule"/>
        <xsd:element ref= "Defeasible_rule"/>
      </xsd:choice>
      <xsd:element ref="Defeasibly_provable" minOccurs="0" maxOccurs="unbounded"/>
      <xsd:element ref="Not_Definitely_provable"/>
      <xsd:element ref="Blocked" minOccurs="0" maxOccurs="unbounded"/>
    </xsd:sequence>
  </xsd:complexType>
</xsd:element>
```

A *Not Defeasibly provable* element consists of the *Atom* in question and its *Not Defeasible proof*. The *Not Defeasible proof* consists of the *Not Definitely provable* element for the *Atom* in question and either a sequence of *Blocked* elements for every rule with head the *Atom* in question and the reason they cannot be applied, or a *Definitely provable* element for the negation of the *Atom*, or by the element *Undefeated* providing an applicable rule r with head the negation of *Atom* in question such that no possibly applicable rule s with head the specified *Atom* is superior to r.

```
<xsd:element name = "Superior">
  <xsd:complexType>
    <xsd:sequence>
      <xsd:element ref="Defeasible_rule"/>
      <xsd:element ref="Defeasibly_provable" minOccurs="0" maxOccurs="unbounded"/>
    </xsd:sequence>
  </xsd:complexType>
</xsd:element>

<xsd:element name = "Not_Defeasibly_provable">
  <xsd:complexType>
    <xsd:sequence>
      <xsd:element ref = "Atom" />
      <xsd:element ref = "Not_Defeasible_Proof" />
    </xsd:sequence>
  </xsd:complexType>
</xsd:element>

<xsd:element name = "Not_Defeasible_Proof">
```

```
<xsd:complexType>
  <xsd:sequence>
    <xsd:element ref= "Not_Definitely_provable"/>
    <xsd:choice>
      <xsd:element ref= "Blocked" minOccurs="0" maxOccurs="unbounded"/>
      <xsd:element ref="Definitely_provable"/>
      <xsd:element ref="Undefeated"/>
    </xsd:choice>
  </xsd:sequence>
</xsd:complexType>
</xsd:element>

<xsd:element name="Undefeated">
  <xsd:complexType>
    <xsd:sequence>
      <xsd:element ref= "Defeasible_rule"/>
      <xsd:element ref="Defeasibly_provable" minOccurs="0" maxOccurs="unbounded"/>
      <xsd:element ref= "Blocked" minOccurs="0" maxOccurs="unbounded"/>
    </xsd:sequence>
  </xsd:complexType>
</xsd:element>
```

3.2 XML Proof Processing

Visualizing a proof explanation requires information about its structure, which must be extracted from the XML document, produced by the system [11], [12] based on the aforementioned XSD Schema. Due to the recursiveness of the xsd, an Recursive Descent Parser (RDP) was implemented. An RDP is a top-down parser built from a set of mutually-recursive procedures (or a non-recursive equivalent) where each such procedure usually implements one of the production rules of the grammar. Thus the structure of the resulting program closely mirrors that of the grammar it recognizes. In our system, the RDP parses the XML document with the assistance of Xerces, the Open Source Apache project's XML parser, and stores the main proof as well as each secondary proof in a different tree–shape structure. Each such structure holds the information required to represent the corresponding proof explanation, i.e. the sequence of rules participating in the proof. For each rule the following information is held:

- the name
- the type (definite or defeasible)
- the head
- the body
- the names of the attacking rules that could defeat it — if such rules exist, or whether the rule is undefeated.

After parsing the XML document and keeping all the appropriate information, the visualization of every proof takes place. In order to visualize every proof, since we consider it has a tree–shape structure, we need to evaluate the height of each node of the proof in the tree. Each node is considered to be either an atom that participates in the body or the head of some rule or the rule itself. For the visualization of the components of the rules , we used the library of [14] which renders each node and the connections between them.

In their approach the digraph representation contains two kinds of nodes:

- literals, represented by rectangles, which they call literal boxes
- rules, represented by circles

Each literal box consists of two adjacent atomic formula boxes, with the upper one of them representing a positive atomic formula and the lower one representing a negated atomic formula.

3.3   Visualization

Definite Proofs. A definite proof is the simplest case of proof explanation. Such a proof consists either of a fact of the literal in question or of a sequence of Strict Rules that fire proving the literal in question.

Not Definite Proofs. A not-definite proof consists of a sequence of proof explanations. Each proof explanation shows why the Strict Rules with head equal to the negation of the literal in question do not fire.

Defeasible Proofs. As mentioned above, a literal is defeasibly provable either if it is already definitely provable or the defeasible part needs to be argued upon. In case the literal is definitely provable, the proof is visualized as described in section 3.3. Otherwise, in order to produce a fully descriptive visualization of a defeasible proof faithful to the reasoning referred to above, several parts are included in the graphic:

1. The main proof, i.e. the sequence of defeasible or definite rules supporting the literal in question.
2. A not definite proof for the negation of the literal in question.
3. A series of not defeasible proofs for every rule attempting to prove the negation of the literal in question. If the specified rule does not fire, the proof consists of a chain of rules supporting a literal in the rule's body that is not provable. Otherwise, if the rule fires but is defeated by a superior applicable counterattacking rule, the proof consists of both the chain of rules proving the negation of the literal and the chain of rules proving the literal in question, and the superiority relation is displayed verbally.

Seperate proofs are visualized in seperate tabs in the graphic.

Not Defeasible Proofs. As described in section 2, a series of conditions must hold in order for a literal to be not defeasibly provable. The visualization of a Non Defeasible Proof consists of two parts:

1. The visualization of the Not Definite Proof for the specified literal.
2. If the literal is not defeasibly provable because all rules that support it do not fire, a Not Defeasible Proof for each such rule is visualized in a seperate tab. If the negation of the literal is definitely provable, then a seperate tab visualizing the Definite Proof is included in the graphic. Otherwise, if there

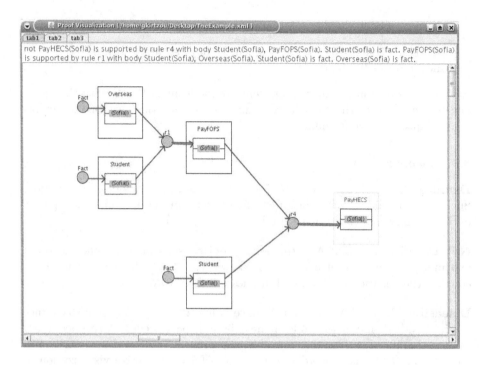

Fig. 1. The main defeasible proof of ∼payHECS(Sofia)

exists an undefeated rule supporting the negation of the literal, the corresponding Defeasible Proof is included in the graphic, and a series of Not Defeasible Proofs for each counterattacking rule with a not defeasibly provable body on separate tabs. Rules that fail to fire because they are defeated by the undefeated rule appear on seperate tabs as well, and the superiority relationship is expressed verbally.

3.4   Example

To demonstrate our tool, we are using as an example a subset of the knowledge base given in [17], modelling part of the Griffith University guidelines on fees. In particular, we consider the following rules:

- r_1: student(X), overseas(X) ⇒payFPOS(X)
- r_2: student(X), overseas(X), exchange(X) ⇒∼ payFPOS(X)
- r_3: student(X) ⇒ payHECS(X)
- r_4: student(X), payFPOS(X) ⇒∼payHECS(X)
- $r_4 > r_3$

The rules represent the following policy:
Overseas students generally pay Overseas Students Fee (FPOS),unless they come from an international exchanged program. All students pay the Higher Education Contribution Scheme (HECS), apart from students who pay FPOS.

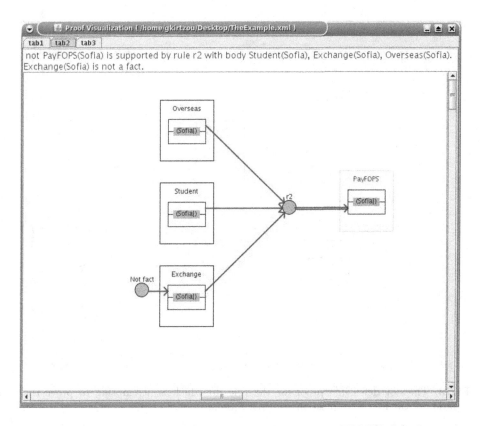

Fig. 2. A not definite proof for the negation of the literal payFPOS(Sofia), due to the recursiveness of defeasible proof

Suppose we have an overseas student named Sofia and we want to query upon whether she has to pay HECS or not. Our knowledge base now comprises of the aforementioned rules in addition to the following facts:

- student(Sofia)
- overseas(Sofia)

$\sim$payHECS(Sofia) is defeasibly provable, since rule r_1 fires establishing the literal payFPOS(Sofia). Therefore, rule r_4 fires supporting payHECS(Sofia). The first tab of the GUI (Figure1) illustrates this sequence of applicable rules. The tree–shaped structure of the proof is achieved with the duplication of the required nodes(literals), leading to a more easy reading form of visualization. Rule r_2 supporting $\sim$payFPOS(Sofia)) does not fire, since Sofia is not an exchange student in our knowledge base. This is demonstrated in the second tab of of our GUI (Figure 2). Rule r_3 supportingpayHECS(Sofia) fires, but loses due to superiority relation. This renders on the third tab of the GUI (Figure 3). The seperate components of this proof are also presented verbally at the top of each tab. As we mentioned in section 3.3 the superiority relationships are presented only verbally.

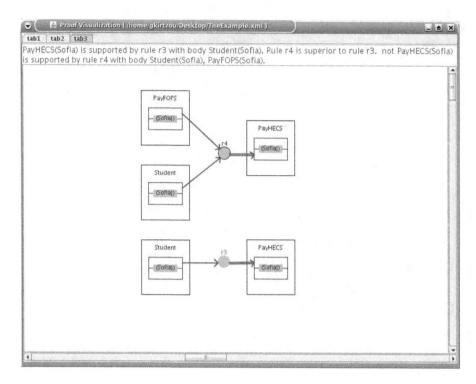

Fig. 3. The not defeasible proof for the rule r_3, which is attempting to prove the negation of the literal in question, payHECS(Sofia)

4   Conclusions and Future Work

This paper attempts to fill the apparent gap in the development of the proof layer of the Semantic Web, by presenting a system for visualizing proofs in the Semantic Web environment. The proposed system uses defeasible logic as the underlying inference system and its adopted proof representation schema is based on enhanced directed graphs that feature a variety of node and connection types for expressing the necessary elements of the defeasible logic proof theory. More specifically, the system offers the capability of visualizing both definite (facts and strict rules only) and non-definite proofs (sequence of proof explanations that show why strict rules with head equal to the negation of the literal in question do not fire) as well as defeasible (either definitely provable or the defeasible part needs to be argued upon) and non-defeasible proofs (not definitely provable plus some additional conditions). Section 2 provides a deeper insight on defeasible logic proof theory. An example was also provided that demonstrates the representational capabilities of the proposed implementation.

As for future directions, there is still room for improvement. The visual representation should incorporate some further elements of defeasible reasoning, like the superiority relationship, which is currently only displayed verbally. Additionally, the proposed software tool should undergo a thorough user evaluation, in order to assess the degree of expressiveness it offers and whether the derived proof visualizations are indeed more comprehensible than the XML-based proofs. An interesting idea would also involve the integration into the system of a visual defeasible theory representation tool, like the one presented in [18]. Then, users would have the ability of (visually) querying the system regarding the proof status of every literal in the rule base and observe a visualization of the corresponding proof trace.

References

1. Grosof, B.N., Horrocks, I., Volz, R., Decker, S.: Description Logic Programs: Combining Logic Programs with Description Logic. In: WWW 2003: Proceedings of the 12th international conference on World Wide Web, pp. 48–57. ACM, New York (2003)
2. Levy, A.Y., Rousset, M.C.: Combining Horn rules and description logics in CARIN. Artificial Intelligence 104(1-2), 165–209 (1998)
3. Rosati, R.: On the decidability and complexity of integrating ontologies and rules. J. Web Sem. 3(1), 61–73 (2005)
4. Sintek, M., Decker, S.: TRIPLE - A Query, Inference, and Transformation Language for the Semantic Web. In: Horrocks, I., Hendler, J. (eds.) ISWC 2002. LNCS, vol. 2342, pp. 364–378. Springer, Heidelberg (2002)
5. Horrocks, I., Patel-Schneider, P.F.: A Proposal for an OWL Rules Language. In: WWW 2004: Proceedings of the 13th international conference on World Wide Web, pp. 723–731. ACM, New York (2004)
6. RuleML: The RuleML Initiative website (2006), http://www.ruleml.org/
7. Antoniou, G., Bikakis, A.: DR-Prolog: A System for Defeasible Reasoning with Rules and Ontologies on the Semantic Web. IEEE Trans. on Knowl. and Data Eng. 19(2), 233–245 (2007)
8. Bassiliades, N., Antoniou, G., Vlahavas, I.: A Defeasible Logic Reasoner for the Semantic Web. International Journal of Semantic Web and Information Systems (IJSWIS) 2(1), 1–41 (2006)
9. Gandhe, M., Finin, T., Grosof, B.: SweetJess: Translating DamlRuleML to Jess. In: International Workshop on Rule Markup Languages for Business Rules on the Semantic Web in conjunction with ISWC 2002, Sardinia, Italy (2002)
10. Eiter, T., Ianni, G., Schindlauer, R., Tompits, H.: Dlvhex: A System for Integrating Multiple Semantics in an Answer-Set Programming Framework. In: Fink, M., Tompits, H., Woltran, S. (eds.) WLP, Technische Universität Wien, Austria. INFSYS Research Report, vol. 1843, pp. 206–210 (2006)
11. Antoniou, G., Bikakis, A., Dimaresis, N., Genetzakis, M., Georgalis, G., Governatori, G., Karouzaki, E., Kazepis, N., Kosmadakis, D., Kritsotakis, M., Lilis, G., Papadogiannakis, A., Pediaditis, P., Terzakis, C., Theodosaki, R., Zeginis, D.: Proof Explanation for the Semantic Web Using Defeasible Logic. In: Zhang, Z., Siekmann, J.H. (eds.) KSEM 2007. LNCS (LNAI), vol. 4798, pp. 186–197. Springer, Heidelberg (2007)

12. Antoniou, G., Bikakis, A., Dimaresis, N., Genetzakis, M., Georgalis, G., Gover-
    natori, G., Karouzaki, E., Kazepis, N., Kosmadakis, D., Kritsotakis, M., Lilis,
    G., Papadogiannakis, A., Pediaditis, P., Terzakis, C., Theodosaki, R., Zeginis, D.:
    Proof explanation for a nonmonotonic Semantic Web rules language. Data Knowl.
    Eng. 64(3), 662–687 (2008)
13. Nute, D.: Defeasible logic. In: Gabbay, D., Hogger, C.J., Robinson, J.A. (eds.)
    Handbook of Logic in Artificial Intelligence and Logic Programming. Nonmono-
    tonic Reasoning and Uncertain Reasoning, vol. 3, pp. 353–395. Oxford University
    Press, Oxford (1994)
14. Kontopoulos, E., Bassiliades, N., Antoniou, G.: Visualizing Defeasible Logic Rules
    for the Semantic Web. In: Mizoguchi, R., Shi, Z., Giunchiglia, F. (eds.) ASWC
    2006. LNCS, vol. 4185, pp. 278–292. Springer, Heidelberg (2006)
15. Antoniou, G., Billington, D., Governatori, G., Maher, M.J.: Embedding Defeasible
    Logic into Logic Programming. Theory Pract. Log. Program. 6(6), 703–735 (2006)
16. Maher, M.J., Rock, A., Antoniou, G., Billington, D., Miller, T.: Efficient Defeasible
    Reasoning Systems. International Journal on Artificial Intelligence Tools 10(4),
    483–501 (2001)
17. Antoniou, G., Billington, D., Maher, M.J.: On the Analysis of Regulations using
    Defeasible Rules. In: HICSS (1999)
18. Bassiliades, N., Kontopoulos, E., Antoniou, G.: A Visual Environment for Devel-
    oping Defeasible Rule Bases for the Semantic Web. In: Adi, A., Stoutenburg, S.,
    Tabet, S. (eds.) RuleML 2005. LNCS, vol. 3791, pp. 172–186. Springer, Heidelberg
    (2005)

Building an Autopoietic Knowledge Structure for Natural Language Conversational Agents

Kiyoshi Nitta

Yahoo! JAPAN Research
Tokyo, Japan

Abstract. This paper proposes a graph structure called an *augmented semantic network* (ASN) which is an extension of the ordinary semantic network (SN). We have developed an experimental conversational agent system and a knowledge structure based on the ASN that can hold a larger number of rules for replying to user utterances and modifying some parts of the rules when necessary. The system operation has shown that the knowledge structure is capable of implementing well-studied conversational models and becoming an autopoietic system. Autopoiesis means the systemic nature of life activity as discussed in the field of life science. An autopoietic system will be able to reproduce its elements as a result of their activity. Although the system will have to be capable of other functional natures to become an autopoietic system, the additional flexibility and extensibility enabled by the ASN-based knowledge structure might be necessary for realizing autopoietic and intelligent conversational agents.

The SN graph structure consists of a vertex set and an edge set whose elements each connect two elements in the vertex set. ASN edges are also able to connect elements in the edge set. The knowledge structure permits concept synthesis by utilizing the ASN's edge modification ability. It removes the restriction on giving meanings from the outside to all concepts in the knowledge structure. Each element of rules represented by the ASN graph structure has a concrete meaning that can be synthesized by other elements of the rules. This capability might further the development of an autopoietic knowledge structure system.

1 Introduction

Turing believed that the intelligence of conversational agents should be judged by their behavior rather than by their implementation technology [1]. This judgment rule has driven developers of conversational agents to use techniques that might allow agents to impress examiners with their human qualities. Those techniques include pronoun replacement, leading topic change, reuse of past user dialogue, and synonym substitution based on static dictionaries [2,3]. Although these have the potential to support machine intelligence in the future, there is no guarantee that they are sufficient to enable it. In this paper, an autopoietic knowledge structure will be discussed in terms of it being a necessary technology for equipping conversational agents with intelligence. While this technology

N. Bassiliades, G. Governatori, and A. Paschke (Eds.): RuleML 2008, LNCS 5321, pp. 211–218, 2008.

may prove to be less than sufficient to enable machine intelligence, it will be an effective part of eventual machine intelligence that is defined by its behavior.

As a basic technology for making knowledge structures into autopoietic systems, we will introduce an augmented semantic network (ASN). This technology permits expressions for modifying the edges of a graph structure. The ordinary graph structure of a semantic network [4] does not permit such expressions. This extension enables complete class-instance semantics across the whole knowledge structure. The global capability of the semantics makes the knowledge structure flexible and extensible. These characteristics may be necessary to achieve autopoiesis in the knowledge structures of conversational agents.

The knowledge structures of conversational agents consist of large numbers of action rules that might be able to reply to user utterances and can modify rules themselves and statuses of the agents. Autopoietic knowledge structures require self-reproducing processes for their rules. By constructing the knowledge structure based on an ASN, a conversational agent can represent the rules as a fully annotated directed graph, which enables implementation of such self-reproducing processes. Although inventing such processes itself is an unresolved problem, the ASN provides a useful basis for exploring the problem.

The rest of this paper is organized as follows. The Augmented Semantic Network section introduces basic technology used to build the knowledge structure. In the Knowledge Structure for Conversational Agents section, we briefly show the knowledge structure implementation as an example application of the augmented semantic network. The characteristics of the ASN are described in the Discussion section. In Related Work, we summarize several relevant studies on conversational agents and semantic networks, and then we conclude this paper in the Conclusion section.

2   Augmented Semantic Network

As a basic model of a knowledge structure that provides flexibility and extensibility, we developed the *augmented semantic network* (ASN). The structure of an ASN is an element set E_{ASN}:

$$E_{ASN} = \{e(e_s, e_d) | e_s, e_d \in E_{ASN} \cup \{\mathbf{null}\}\} \tag{1}$$

The notation e means an *element*. The element e has the binomial structure (e_s, e_d). Each of the elements e_s, e_d is either an element of the set E_{ASN}, or an empty element **null**. The element $e(e_s, e_d)$ is called an *edge* if both e_s, e_d are elements of E_{ASN}. The element e is otherwise called a *vertex*. The element $e(e_s, e_d)$ means a relation from the element e_s to the element e_d.

The ASN structure is a variation of extensions of the semantic network (SN). The structure of an ordinary SN is a directed graph:

$$SN = (V_{SN}, E_{SN}) \tag{2}$$

$$V_{SN} = \{v | v \text{ is a vertex}\} \tag{3}$$

$$E_{SN} = \{e(v_s, v_d) | v_s, v_d \in V_{SN}, e \text{ is an edge}\} \tag{4}$$

The set V_{SN} contains vertices, and E_{SN} contains edges. In the upper side of Figure 1, vertices v_s, v_d are shown as circles, and an edge e is shown as an arrow from v_s to v_d. The subset of ASN structure ASN_{VE} has the same expressive power as the SN structure when element sets $E_v, E_e \subset E_{ASN}$ are defined as follows:

$$ASN_{VE} = (E_v, E_e) \tag{5}$$
$$E_v = \{e(e_s, e_d)|e_s, e_d = \mathbf{null}\} \tag{6}$$
$$E_e = \{e(e_s, e_d)|e_s, e_d \in E_v\} \tag{7}$$

The set E_v contains vertices, and E_e contains edges. The difference between ASN and SN ($= ASN_{VE}$) is whether the restriction $e_s, e_d \in E_v$ in Equation (7) exists. That means the ASN has an additional ability by which an ASN edge can connect between edges in the same way as between vertices. The edge can be shown as an arrow with one of its terminal points connected to the intermediate point of a line segment. The ASN element e_i is represented by an arrow which connects vertex e_c and edge e_e in the lower part of Figure 1. Edges like e_i can only exist in an ASN.

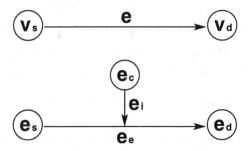

Fig. 1. Semantic Network (upper) and Augmented Semantic Network (lower)

The semantics of the ASN are also inherited from those of the SN. In the SN semantics, vertices mean concepts or facts, and edges means relations between vertices. The concrete meanings of vertices and an edge are defined by labels, which come from outside of the SN structure. The ASN semantics also include the semantics of edge elements which connect edge elements. Edge elements are relations between elements which can be either vertices or edges in the ASN semantics much like in the SN semantics. These additional semantics enable another method for defining concrete meanings.

3 Knowledge Structure for Conversational Agents

An experimental version of this conversational agent being developed is equipped with a knowledge structure based on the augmented semantic network (ASN). The agent is coded in the C++ programming language. The code size is about

23,000 lines. The agent performs text message conversations with human users by applying a *dialogue script tracer* to *dialogue script rules*, which are expressed on the ASN. The dialogue script tracer is a part of the agent. Its single running instance has a contextual position pointer. It changes the contextual position by following ASN graph elements of the dialogue script rules and interpreting user utterances. All elements of the dialogue script rules are fully annotated by classes representing clear meanings for the dialogue script tracer. These annotations are represented by ASN edges which connect from class vertices to instance rule elements. These classes are defined on an extensible basic framework, which is also expressed on the ASN.

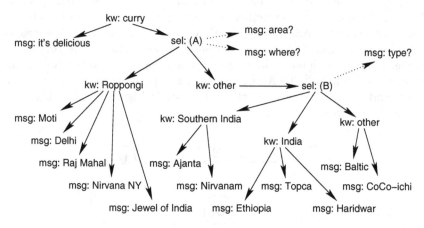

Fig. 2. Example of a dialogue script rule

Figure 2 is a simple example diagram of the dialogue script rule. This example rule contains five classes of instance elements. Three classes of vertex elements are expressed as text labels, each of which has a class prefix. Vertices prefixed by 'kw', 'msg', and 'sel' mean trigger keywords, reply messages, and selection branches, respectively. Two classes of edge elements are expressed as arrows in the diagram, where solid arrows mean possible script directions, and dotted arrows mean reply messages for selection branches. When the dialogue script tracer tries to process an element of the dialogue script rule, it investigates the class of the element and finds an appropriate function by the class.

4   Discussion

The characteristics of SN and ASN are compared in Table 1. This table shows how the methods differ when providing semantics to graph structures. The meanings of SN relations are given from outside of the SN. The meanings of ASN relations are given by class vertices which are connected by instance relations. However, the meanings of these vertices are given from outside of the ASN. While the final semantic resolution comes from outside in both methods, the ASN method has several advantages. Because one ASN relation may belong to more than one semantic category, there is no need for any two vertices to have multiple edge entities. This increases the space efficiency of graph repositories and reduces the access cost for related information. In addition, because the meanings of ASN relations can be synthesized using concept vertices, even if the given concept sets are the same, an ASN is more able to express conversational agent knowledge.

Table 1. Comparison of SN and ASN

SN	ASN
give meanings of relations directly from the outside	give meanings of relations through instance relations from vertices
one relation must belong to one semantic category	one relation may belong to multiple semantic categories
relation semantics fixed	relation semantics can be synthesized
free expression of concept attributes	free expression of concept relations

Minsky [5] depicted ideal mechanisms for the human brain to enable unique traits like emotions, thinking, and resourcefulness. Those mechanisms included the six-level model of mind, the critic-selector model, and the panalogy architecture. All of these mechanisms require multiple qualification of a knowledge entity which is processed by them. Unlike an SN, an ASN can qualify edges not only by giving a name, but also through some combination of ASN graph elements. This ability is listed in Table 1 as synthesized semantics. This means that the knowledge represented by an ASN may include unnamed concepts corresponding to the qualification edges. Presumably, this feature will greatly help to implement such mechanisms.

It is meaningful that dialogue script rules are distributed widely through the Internet. This might help to get past the real world modeling bottleneck by aggregating small individual efforts. Although we focused on the extensibility of conversational agent knowledge structures by using the ASN, describing such rules in a standardized rule markup language will be a good solution and an important future task.

5   Related Work

Many conversational agents have been developed so far. The earliest conversational agent, Weizenbaum's ELIZA [2], was reported in 1966. It was equipped with a reply function that repeated user messages by substituting pronouns for adequate pronouns, and thus acted as a psychiatrist. To some degree, it gave users the impression they were conversing with a human psychiatrist. Colby's PARRY [6] acted as a psychiatric patient, and it had no need to reply with correct answers for its users. This showed that it is possible to implement a conversational agent which cannot be distinguished from a human psychiatric patient. The CONVERSE agent [6] integrated a commercial parser, a query interface to access a parse tree, large-scale natural language resources like Wordnet, and other natural language processing technologies commonly available in 1997. It achieved a high percentage of completion, and won an annual medal of the 1997 Loebner prize [7]. Wallace's ALICE[1] has promoted the XML-based rule description language AIML. Wallace developed the AIML processor and the ALICE subset knowledge data in the AIML format as open source resources. Several conversational agents [8,9] have been developed based on these ALICE resources. ALICE has won three annual medals of the Loebner prize.

The above agents use typed natural language text as an interacting medium with human users. The steady growth of computational power has enabled the implementation of several new media technologies, though, that could be useful for this purpose. These include computer controlled visible characters [10,11], voice recognition and synthesis, and eye and gesture trackers.

Many studies have focused on the human-assisting usage of natural language conversation technologies. Allen et al. developed the TRIPS library [3], which can be used to add a natural language conversation capability to interactive systems. TRIPS has a framework that can be used to model any conversational task, and TRIPS-integrated interactive systems have an autonomic prompting capability to facilitate primary interactive tasks for human users. Similar studies have focused on information retrieval tasks [12,13,14], question and answering tasks [15], and home electronic appliance controlling tasks [16].

Although the implemented conversational agents offer rule generating functions, they require the user to provide explicit instructions for rule element modifications. These functions can maintain high quality rules but require the user to have a rich understanding of the knowledge structure. Case-based learning technologies [17,18] are likely to be needed to enable more users of this agent. A recent methodology of note is the harnessing of the cooperative efforts of huge numbers of Internet users, which is one form of the concept referred to as Web 2.0. Increasing the number of potential users is important to enlarge the amount of conversational rule knowledge accumulated through Web 2.0 methodology.

Most of the above conversational agents have their own conversational models. While many of these models were designed to handle conversations between a user and an agent, Knott & Vlugter [19] implemented other types of models to handle conversations between multiple speakers.

Semantic networks have been used for a long time for several purposes in computing, such as natural language processing, logic programming, and database access interfaces. Semantic networks have been so widely used, in fact, that there was no commonly accepted strict SN definition. However, Griffith [4] defined it universally by dividing the net into an abstract net and a conveying net. Our SN extension should be understood in terms of the conveying semantic network, which puts much of the emphasis on its implementation methodology. The SN visibility and computability was important in past applications, but flexibility, extensibility, and efficiency are more important for autopoietic knowledge structures.

---

[1] http://alicebot.org/articles/wallace/dont.html

6   Conclusion

We have proposed an augmented semantic network (ASN) as a basis for constructing autopoietic knowledge structure systems. The ASN graph structure has a unique capability in having edge elements that can modify other edge elements. Through this capability, class—instance semantics can be introduced to design and implement the ASN-based knowledge structure. Such semantics provide conversational rules, which are the main part of our conversational agent knowledge structure, and simple directed graph expressions. These rules can utilize various conversational models already developed, define new meanings by synthesizing existing definitions, and reproduce new variation of rules. This structure satisfies the flexibility and extensibility requirements for autopoietic systems.

The next steps in this study will be to perform further experiments on rule reproducing functions, construct more complex conversational models, integrate information retrieval and logical inference technologies, and develop evaluation methodologies. We believe that such work will be a necessary step to realize an autopoietic knowledge structure for conversational agents equipped with human-like intelligence.

References

1. Turing, A.M.: Computing machinery and intelligence. Mind 59(236), 433–460 (1950)
2. Weizenbaum, J.: ELIZA — a computer program for the study of natural language communication between man and machine. Commun. ACM 9(1), 36–45 (1966)
3. Allen, J., Ferguson, G., Stent, A.: An architecture for more realistic conversational systems. In: IUI 2001: Proceedings of the 6th international conference on Intelligent user interfaces, pp. 1–8. ACM Press, New York (2001)
4. Griffith, R.L.: Three principles of representation for semantic networks. ACM Trans. Database Syst. 7(3), 417–442 (1982)
5. Minsky, M.: The Emotion Machine: Commonsense Thinking, Artificial Intelligence, and the Future of the Human Mind. Simon & Schuster (2006)
6. Wilks, Y.: Machine Conversations. Kluwer Academic Publishers, Norwell (1999)
7. Loebner, H.G.: In response. Commun. ACM 37(6), 79–82 (1994)
8. Galvao, A.M., Barros, F.A., Neves, A.M.M., Ramalho, G.L.: Persona-aiml: An architecture developing chatterbots with personality. In: AAMAS 2004: Proceedings of the Third International Joint Conference on Autonomous Agents and Multiagent Systems. IEEE Computer Society, Los Alamitos (2004)
9. Sing, G.O., Wong, K.W., Fung, C.C., Depickere, A.: Towards a more natural and intelligent interface with embodied conversation agent. In: CyberGames 2006: Proceedings of the 2006 international conference on Game research and development, pp. 177–183. Murdoch University, Australia (2006)
10. Cassell, J., Bickmore, T., Billinghurst, M., Campbell, L., Chang, K., Vilhjálmsson, H., Yan, H.: Embodiment in conversational interfaces: Rea. In: CHI 1999 Proceedings of the SIGCHI conference on Human factors in computing systems, pp. 520–527. ACM Press, New York (1999)

11. Thorisson, K.R.: Gandalf: an embodied humanoid capable of real-time multimodal dialogue with people. In: AGENTS 1997: Proceedings of the first international conference on Autonomous agents, pp. 536–537. ACM Press, New York (1997)
12. Chen, L., Sycara, K.: Webmate: a personal agent for browsing and searching. In: AGENTS 1998: Proceedings of the second international conference on Autonomous agents, pp. 132–139. ACM Press, New York (1998)
13. Luke, S., Spector, L., Rager, D., Hendler, J.: Ontology-based web agents. In: AGENTS 1997: Proceedings of the first international conference on Autonomous agents, pp. 59–66. ACM Press, New York (1997)
14. Etzioni, O., Weld, D.: A softbot-based interface to the internet. Commun. ACM 37(7), 72–76 (1994)
15. Kiyota, Y., Kurohashi, S., Kido, F.: "dialog navigator": a question answering system based on large text knowledge base. In: Proceedings of the 19th international conference on Computational linguistics, Morristown, NJ, USA, pp. 1–7. Association for Computational Linguistics (2002)
16. Yates, A., Etzioni, O., Weld, D.: A reliable natural language interface to household appliances. In: IUI 2003: Proceedings of the 8th international conference on Intelligent user interfaces, pp. 189–196. ACM Press, New York (2003)
17. Liu, H., Lieberman, H.: Programmatic semantics for natural language interfaces. In: CHI 2005 extended abstracts on Human factors in computing systems, pp. 1597–1600. ACM Press, New York (2005)
18. Bauer, M., Dengler, D., Paul, G.: Instructible information agents for web mining. In: IUI 2000: Proceedings of the 5th international conference on Intelligent user interfaces, pp. 21–28. ACM Press, New York (2000)
19. Knott, A., Vlugter, P.: Multi-agent human–machine dialogue: issues in dialogue management and referring expression semantics. Artif. Intell. 172(2-3), 69–102 (2008)

A Functional Spreadsheet Framework for Authoring Logic Implication Rules

Marcelo Tallis[1] and Robert M. Balzer[2]

[1] University of Southern California / Information Sciences Institute
4676 Admiralty Way, Suite 1001, Marina del Rey, CA 90292, U.S.A.
[2] Teknowledge Corp., 2595 E. Bayshore Road, Suite 250,
Palo Alto, CA 94303, U.S.A.
tallis@isi.edu, bbalzer@teknowledge.com

Abstract. This paper introduces a functional spreadsheet framework for authoring logic implication rules. This framework was conceived with the objective of
reproducing many of the characteristics that make spreadsheet programming
accessible to end-users. In the proposed framework, rule authors describe the
semantics of a binary relation by constructing a functional spreadsheet model
that computes the image of that binary relation. This model is subsequently
translated into a collection of logic implication rules. We implemented and integrated this framework into a deductive spreadsheet system that extends Microsoft Excel with the World Wide Web Consortium (W3C) standard ontology
language OWL + SWRL.

Keywords: Deductive Spreadsheets. End-users programming. Authoring Logic
Implication rules. SWRL.

1 Introduction

The spreadsheet is one of the most widespread used computing applications. It has
been estimated that there are over 55 million spreadsheet users in the U.S. alone [1].
These end-users succeed in writing spreadsheet programs for getting their job done
albeit the vast majority of these users are not trained programmers.

From a computational perspective, spreadsheets are functional programs in which
spreadsheet cells are used as variables. However, the intuitiveness of the spreadsheet
development largely hides the extent to which it is actually a programming activity.
[2] identified as contributors to the spreadsheet success the immediate feedback
through formula evaluation, the tabular grid format which leads to a natural representation for many problems, the possibility of reducing complexity by splitting formulas
over different cells, and the rather declarative nature of most aspects of spreadsheet
languages. Because the current values of the variables are permanently displayed, a
user can quickly see the effects of changes to input values or formulas, encouraging
an explorative style of programming [3].

Conventional spreadsheet applications are good at tackling numerically intensive
problems. However, many high-level problem-solving tasks are better fitted to logical
reasoning methods. In order to tackle these kinds of problems, logic deductive capabilities are needed.

N. Bassiliades, G. Governatori, and A. Paschke (Eds.): RuleML 2008, LNCS 5321, pp. 219–226, 2008.
© Springer-Verlag Berlin Heidelberg 2008

We have developed a system that extends the spreadsheet paradigm with deductive reasoning capabilities [4]. This system extended Microsoft Excel with the OWL and SWRL logic formalism [5]. In this *deductive spreadsheet* system, spreadsheet models can be augmented with logic rules and ontological definitions. Based on these definitions, users can build spreadsheet models that blend seamlessly logic assertions and inferences with conventional spreadsheet values and formulas. This system is described in more detail in [4]. The purpose of this paper is to describe our framework for authoring the logic implication rules needed to reason in a problem domain. These rules are of the form of an implication between an antecedent and consequent. The intended meaning can be read as: whenever the conditions specified in the antecedent hold, then the conditions specified in the consequent must also hold. The following is an example of a logic implication rule:

```
Sibling(?x, ?y) ^ Gender(?y, "Female") → Sister(?x, ?y)
```

The design of this framework was driven by the objective of incorporating many of the characteristics mentioned above attributed with making spreadsheet programming accessible to end-users.

Our design was inspired by the work of [6]. In [6] the authors proposed a framework that integrated user-defined functions into the spreadsheet grid. In their framework, user defined functions are defined in terms of formulae that are contained in a specified part of the spreadsheet. The function signature is defined by allocating some subset of the cells as function parameters.

Our rule authoring framework adopted the idea of having logic rules be defined in terms of formulae that are contained in a specified part of the spreadsheet. However, in our framework these formulae rely on special spreadsheet operators conceived for this particular purpose. These rule definition models are called *Property Definition Worksheets* because they describe sufficient conditions for defining an ontology property. Property Definition Worksheets (PDWs) are conceived as the means for authoring logic rules within the spreadsheet paradigm. These PDWs are automatically and invisibly translated into conventional SWRL rules and loaded into the deductive logic engine to compute the defined properties.

2   Deductive Spreadsheets

Fig. 1 shows a table that might appear in a typical spreadsheet. This table lists a group of persons where each row describes a person and each column indicates the values for one of the person's properties (e.g., gender). Some values in the table can be deduced from other values in the same table. For example, the siblings of a person can be deduced from the father and the mother properties of that person and of other persons in the table. Our objective was to enrich Excel with deductive logic capabilities for automatically computing values like the sibling property. We provided this capability while preserving the mechanics of interacting with spreadsheets to which users are already accustomed: for example, entering data in some cells of a spreadsheet (e.g., the Mother column) and watching that data propagate to other cells of the spreadsheet (e.g., the Sibling column).

2.1 Deductive Logic Spreadsheet Architecture

At the highest level, the integration of a deductive logic engine into a spreadsheet requires that information from the spreadsheet be available as facts or *inputs* to the logic engine and that the deductions (or inferences) it draws from those facts can be inserted as derived values or *outputs* in the spreadsheet.

	A	B	C	D	E	F	G
1	Person						
2							
3	FirstName	LastName	Gender	Mother	Father	Sibling	Sister
4	Abraham	Simpson	Male				
5	Mona	Simpson	Female				
6	Clancy	Bouvier	Male				
7	Jacqueline	Bouvier	Female				
8	Homer	Simpson	Male	Mona Simpson	Abraham Simpson		
9	Marge	Simpson	Female	Jacqueline Bouvier	Clancy Bouvier	Patty Bouvier	Patty Bouvier
10						Selma Bouvier	Selma Bouvier
11	Patty	Bouvier	Female	Jacqueline Bouvier	Clancy Bouvier	Marge Simpson	Marge Simpson
12						Selma Bouvier	Selma Bouvier
13	Selma	Bouvier	Female	Jacqueline Bouvier	Clancy Bouvier	Marge Simpson	Marge Simpson
14						Patty Bouvier	Patty Bouvier
15	Bart	Simpson	Male	Marge Simpson	Homer Simpson	Lisa Simpson	Lisa Simpson
16						Maggie Simpson	Maggie Simpson

Fig. 1. A typical spreadsheet with deductive capabilities

Since the deductive logic engine and the spreadsheet have very different notions of the primitive data on which they operate (respectively triples and cells), a mapping must be established between the two so that data can be extracted from and placed into the spreadsheet. The OWL deductive logic engine operates on (subject, property, object) triples indicating that "a property of subject is object." Thus, (A8:G8, Gender, C8) or (A8:G8, Father, E8) respectively indicate that the gender of A8:G8 is C8 and that the father of A8:G8 is E8. The third element of a triple is the value contained in a spreadsheet cell (for inputs to the deductive logic engine) or the value placed in a spreadsheet cell (for outputs from the deductive logic engine). The subject and property for the triple are specified for each cell that participates as input for, or output from, the deductive logic engine and these declarations constitute the mapping between spreadsheet values and the deductive logic engine's knowledgebase.

An ontology and set of rules are also loaded into the deductive logic knowledgebase. The ontology defines the set of terms used to describe the types and properties used in the rules. The ontology and rules can be loaded from an external OWL document or they can be defined within a Deductive Spreadsheet. The definition of logical rules is the main focus of this paper and will be discussed in the next section.

There are different types of deductive logic mapping specifications. Some of them affect an entire area of a spreadsheet, like the *Instance Table declaration*. Others affect only a single cell, like the *Cell-Mapping declaration*.

2.2 Instance Tables

Many of the cells that contain a deductive logic mapping are part of a table whose rows represent instances of a specific type and each column represents a property of the type of instance contained in the table (like the example table in Fig. 1). This conventional way of representing data about instances is exploited to provide an aggregated mechanism to specify deductive logic mappings. In this mode of deductive logic mapping specification, mappings declarations are entered and displayed only in the cells corresponding to the column headings of the instances table. Input mappings are entered as "<=Property(SubjectType)" and output mappings are entered as "=>Property(SubjectType)". For example, cell C3 of the table in Fig. 1 contains the following mapping declaration:

```
<=Gender(Person)
```

All column mappings of a table should agree in the subject type. We call this kind of mapping specification an *Instance Table declaration*.

2.3 Cell-Mapping Declarations

Unlike *instance tables*, *cell-mapping* declarations are confined to a single cell. Alike *instance tables* mappings, there exist both input and output varieties. *Input cell-mapping declarations* reflect the content of spreadsheet cells as facts in the KB. An *input cell-mapping* has the form: **<=P(SubjectRange)** where *P* is the name of a property and *SubjectRange* refers to a spreadsheet range that defines or contains an instance reference. The *input cell-mapping declaration* maps the instance *S* referenced by *SubjectRange* and the content *O* of the cell that holds the cell mapping rule into the OWL triple (*S, P, O*).

Output cell-mapping declarations can be viewed as queries that fill in spreadsheet cells with values retrieved from the KB. An output Cell-mapping declaration has the form: *=>P(SubjectRange)* where *P* is the name of a property and *SubjectRange* refers to a spreadsheet range that defines or contains an instance reference.

The semantics of an output cell-mapping declaration is as follows: a cell-mapping declaration *r* = "*=>P(SubjectRange)*" yields the set $O = \{O_1, O_2,..., O_n\}$ such that (*S, P, O_i*) is a triple entailed by the KB and *S* is the instance referenced by *SubjectRange*, for $1 <= i <= n$. If *O* has only one member then that value is stored in the cell that holds the cell-mapping declaration. In this case *r* is said to be *single-valued*. If O has more than one member then one value is stored in the cell holding the mapping declaration and the remaining values are stored, one value per cell, in the rows underneath the row that contains the mapping declaration. The values stored underneath the mapping rule are called *details* and *r* is said to be *multi-valued*. In this case, additional spreadsheet rows will automatically be inserted if they are needed to avoid the details of the mapping rule to spillover preexisting spreadsheet content.

Fig. 2 a) shows an example of an output cell-mapping declaration. In this example, the user entered the declaration "=>Sibling(A1)" in B1. The cell A1 contains a reference to **Bart Simpson**. Hence, this declaration produces two values: **Lisa Simpson** and **Maggie Simpson.** The first value, **Lisa Simpson,** fills the cell that holds the mapping declaration. The second value, **Maggie Simpson,** constitutes a *detail* row and it is inserted in B2, one row underneath the row containing the mapping declaration.

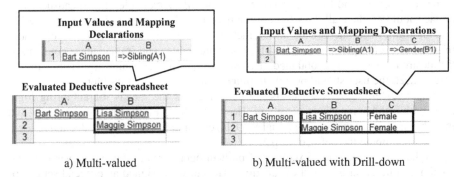

a) Multi-valued   b) Multi-valued with Drill-down

Fig. 2. Output Cell-Mapping Declarations. The top grid shows the values and mapping declarations as they are entered by the user. The bottom grid corresponds to the evaluation of the mapping declarations of the top grid. The boxed area in the bottom grid corresponds to the result of the mapping declaration.

Cells holding output cell-mapping declarations can also be referenced by spreadsheet formulas and other output cell-mapping declarations. A particular situation occurs when a cell-mapping declaration that is referred by other mappings or spreadsheet formulas is multi-valued. We handled this situation by automatically copying dependent mappings and formulas into the detail rows of the multi-valued mapping. Fig. 2 b) illustrates this case.

The following section describes how the logic rules that support logical inferences are defined.

3 Property Definition Worksheets

Property Definition Worksheets (PDWs) is a framework for authoring spreadsheet models that define logic implication rules. These rules are of the form of an implication between an antecedent and consequent. The intended meaning can be read as: whenever the conditions specified in the antecedent hold, then the conditions specified in the consequent must also hold. The following is an example of a logic implication rule:

```
Sibling(?x, ?y) ^ Gender(?y, "Female") → Sister(?x, ?y)
```

These rule definition models are called *Property Definition Worksheets* because they describe sufficient conditions for defining an ontology property. In the above example, the defined property was **Sister**. The design of this framework was driven by the objective of incorporating many of the characteristics attributed with making spreadsheet programming accessible to end-users.

A PDW is a spreadsheet model that a user builds to define an ontology property. Each PDW is implemented as an Excel worksheet. For the purpose of this model, a property is regarded as a function that maps an instance of the property domain into a corresponding set of instances or data values of the property range.

A PDW allocates a cell to hold the *input* parameter to this property-derived function. The task of the user is to author the set of formulae that produce the set of *output* values that correspond to the PDW input. In the **Sister** example, the spreadsheet formulae should produce all the sisters of the instance of person stored in the *input* cell.

This framework encourages an interactive and incremental mode of development in which a user experiments with different input values and immediately confirms if the desired results were obtained. Because the formula results are visible in the spreadsheet, the user should be able to pinpoint and correct any problems.

Property Definition Worksheets are conceived as the means for authoring logic rules within the spreadsheet paradigm. However, these PDWs are automatically and invisibly translated into conventional SWRL rules and loaded into the deductive logic engine to compute the defined properties.

We have found that the Excel set of built-in operators and functions was not adequate for computing this kind of functions. Hence, we designed an especial set of operators for this purpose in particular. These operators specialize in propagating and constraining instances and values related by ontology properties and classes.

We modeled the PDW operators after a feature found in Excel: Data range filters. Excel offers an Advanced Filter command that can be used to select data that satisfy complex criteria. This command requires that the data range to be filtered contains column labels. The criteria are specified in a separate range of the worksheet and must contain column labels too. The filter criteria are specified by aligning data patterns to be matched with the corresponding criteria columns. The patterns to be matched are arranged in one or more rows. All the patterns contained in the same row are conjunctive and the different rows are disjunctive.

The PDW operators that we designed are similar to the Excel Advanced Filter command. And alike the filter command, they operate on table operands that we call *operand tables*.

Fig. 3 a) illustrates our Filter operator (at A29). A Filter operator has two arguments, a source operand table and a criteria operand table. A Filter operator fills the content of the result operand table with the source data that satisfies the criteria conditions. The source of the Filter expression of Fig. 3 a) is the **persons** operand table (located at A9:C21), and its criteria is the **criteria** operand table (located at A24:C26). Its result automatically fills the **Siblings** operand table (located at A29:A32).

Another PDW operator that we introduced is the *Load* operator (Fig. 3 a) at A9). The *Load* operator takes as a parameter a class name and fills the operand table with the instances of that class. For the *Load* operator the column labels should match property names. In this particular example the **Persons** operand table will be filled with the instances of the class "Person", their Fathers and their Mothers.

Operand tables are data structures that represent the operands and results of PDW operators. An operand table consists of a spreadsheet range containing a *table label* in its first row, one or more *column labels* in its second row, and *table content* from its third row until the end of the table. An operand table can optionally include a *table expression* in its first row next to its table label. If an operand table includes a table expression then its content is computed automatically by that expression. In the other hand, if the operand table doesn't include a table expression then its content should be entered manually and it should consist of literal values, mapping declarations, or spreadsheet formulas. When the table content is entered manually, usually the mapping declarations or spreadsheet formulas it contains depend on the value of the PDW input cell.

a) Evaluation for input **Bart Simpson**

	A	B	C
1	Property Name:	Sibling	
2			
3	Input Instance Placeholde	Bart Simpson	
4			
5	Result Table Name:	Siblings	
6	Result Column Name:	Person	
7			
8			
9	Persons:= Load(Person)		
10	Person	Father	Mother
11	Abraham Simpson		
12	Bart Simpson	Homer Simpson	Marge Simpson
13	Clancy Bouvier		
14	Homer Simpson	Abraham Simpson	Mona Simpson
15	Jacqueline Bouvier		
16	Lisa Simpson	Homer Simpson	Marge Simpson
17	Maggie Simpson	Homer Simpson	Marge Simpson
18	Marge Simpson	Clancy Bouvier	Jacqueline Bouvier
19	Mona Simpson		
20	Patty Bouvier	Clancy Bouvier	Jacqueline Bouvier
21	Selma Bouvier	Clancy Bouvier	Jacqueline Bouvier
22			
23			
24	Criteria:=		
25	Father	Mother	Person
26	Homer Simpson	Marge Simpson	<> Bart Simpson
27			
28			
29	Siblings:= Filter(Persons,Criteria)		
30	Person		
31	Lisa Simpson		
32	Maggie Simpson		
33			

H ◀ ▶ H \ Sister \ Sibling / Child / Descende | ◀ |

b) Input Values and Mapping Declarations only

	A	B	C
1	Property Name:	Sibling	
2			
3	Input Instance Placeholder:		
4			
5	Result Table Name:	Siblings	
6	Result Column Name:	Person	
7			
8			
9	Persons:= Load(Person)		
10	Person	Father	Mother
11			
12			
13			
14			
15			
16			
17			
18			
19			
20			
21			
22			
23			
24	Criteria:=		
25	Father	Mother	Person
26	=>Father(B3)	=>Mother(B3)	=notequal(B3)
27			
28			
29	Siblings:= Filter(Persons,Criteria)		
30	Person		
31			
32			
33			

H ◀ ▶ H \ Sister \ Sibling / Child / Descende | ◀ |

Fig. 3. Sibling PDW

Fig. 3 shows the PDW corresponding to the definition of the **Sibling** property. The worksheet in the left side shows the PDW evaluated for the input **Bart Simpson.** The worksheet in the right side for illustrative purposes shows only the values and the formulas entered by the user with not evaluation. The PDW input cell is B3. And the output range, indicated by a pair *(table label, column label),* is indicated in B5:B6 and points to the **person** column of the **Siblings** operand table.

The strategy used to compute the siblings of the input is to filter the persons having the same father and mother than the input but that are different than the input itself. The **Persons** table is filled by the Load(Person) operator (at A9) with the instances of the class "Person", their Fathers and their Mothers. For the **Criteria** operand table the content of the **Father** column is filled by the "=>Father(B3)" mapping which will insert the Father of the input instance. The content of the **Mother** column is filled by the "=>Mother(B3)" mapping which will insert the Mother of the input instance. Finally, the content of the **Person** column is populated with the "=notEqual(B3)" formula which will insert a Different Than input instance condition. Finally, the **Siblings** operand table is filled with a Filter(**Persons,Criteria**) expression which selects all Person instances whose Father matches the Father of the input instance, whose Mother matches the Mother of the input instance, and that are different than the input instance. This table constitutes the output of the **Sibling** property definition worksheet.

The **Sibling** PDW generates the following SWRL rule:

```
Mother(?input,?M) ^ Father(?input,?F) ^ Person(?P) ^ Fa-
ther(?P,?F) ^ Mother(?P,?M) ^ differentIndividualsA-
tom(?P,?input) → Sibling(?input,?P)
```

4   Related Work

There have been several attempts for integrating logic deductive capabilities into spreadsheet applications. In most of these systems, like [7], logic rules still have to be expressed textually using a logic formalism. Rule authors do not perceive any substantial difference with respect to conventional rule editors. In [8] the author propose a different approach for integrating symbolic reasoning into spreadsheets. He proposes to couple Excel with a Datalog-like framework. This framework is better suited for treating relations (or tables) as first-class objects that can be operated upon. In this framework, it is possible to define a table as a formula involving other tables and it lends well at reducing the complexity of formulas by defining intermediate tables.

Acknowledgments. Our thanks to all anonymous reviewers. This work was supported by DARPA under agreement W31P4Q-05-C-R069.

References

1. Boehm, B., Horowitz, E., Madachy, R., Reifer, D., Clark, B.K., Steece, B., Winsor Brown, A., Chulani, S., Abts, C.: Software Cost Estimation with COCOMO II. Prentice Hall, Englewood Cliffs (2000)
2. Nardi, B., Miller, J.: An Ethnographic Study of Distributed Problem Solving in Spreadsheet Development. In: Proc. CSCW, October 1990, pp. 197–208. ACM, New York (1990)
3. Spenke, M., Beilken, C.: A Spreadsheet Interface for Logic Programming. In: Proceedings of the ACM CHI 1989, Austin, Texas, April 30 - June 4, 1989, pp. 75–80 (1989)
4. Tallis, M., Waltzman, R., Balzer, R.: Adding deductive logic to a COTS spreadsheet. The Knowledge Engineering Review 22(03), 255–268 (2007)
5. Horrocks, I., Patel-Schneider, P.F., Boley, H., Tabet, S., Grosof, B., Dean, M.: SWRL:A Semantic Web Rule Language Combining OWL and RuleML (2004),
   http://www.w3.org/Submission/SWRL/
6. Peyton Jones, S., Blackwell, A., Burnett, M.: A user-centred approach to functions in Excel. In: Proceedings of the 8th ACM SIGPLAN, Uppsala, Sweden, pp. 165–176 (2003)
7. Valente, A., Van Brackle, D., Chalupsky, H., Edwards, G.: Implementing logic spreadsheets in LESS. The Knowledge Engineering Review 22, 237–253 (2007)
8. Cervesato, I.: NEXCEL, a deductive spreadsheet. The Knowledge Engineering Review 22(03), 221–236 (2007)

Please Pass the Rules: A Rule Interchange Demonstration

Gary Hallmark [1], Christian de Sainte Marie[2], Marcos Didonet Del Fabro[2],
Patrick Albert [2], and Adrian Paschke[3]

[1] Oracle
gary_hallmark@yahoo.com
[2] ILog
{csma,mddfabro,palbert}@ilog.fr
[3] Free University Berlin
adrian.paschke@gmx.de

Abstract. It is commonly accepted that separating the declarative rules from the procedural code of an application makes that application easier to understand and easier to modify. What has been lacking is a standard representation for rules – until now. Using the W3C's Rule Interchange Format [1], rules will be exchanged and revised among three different rule systems: ILOG JRules [2], Oracle [3], and Prova [4].

Keywords: Rule Interchange, Web Rules, General Rule Markup Language, Rule Standard Language.

1 A Typical Use Case

The W3C RIF Use Cases and Requirements [12] first use case is about negotiating business contracts across rule platforms. John's company, Fragrant Fish, Inc, uses Oracle Business Rules (OBR) [3] in a Service Oriented Architecture to manage his supply chain. Having declaratively "exorcised" his business rules from his application logic, John wants to negotiate the rules with his suppliers. He picks his most important rule to propose to his suppliers:

```
if an item is perishable and
    it is delivered to John more than 10 days
    after the scheduled delivery date

then the item will be rejected
```

One supplier, Jane, uses ILOG rules [2] and another supplier, Jose, uses Prova rules in a Prova Web inference service [4] deployed on a Rule Responder Enterprise Service Bus (ESB) [5]. John uses a "Core" subset of the RIF Basic Logic Dialect [6] to express his proposed rule. This subset can be interchanged between production rule systems (like ILOG and Oracle) and logical rule systems (like Prova).

John's rule can be expressed in RIF-BLD presentation syntax as follows:

```
Prefix(ff http://fragrantfish.com/biz-vocabulary#)

Forall ?item ?deliverydate ?scheduledate (
    ?item[ff:status -> "rejected"] :-
```

N. Bassiliades, G. Governatori, and A. Paschke (Eds.): RuleML 2008, LNCS 5321, pp. 227–235, 2008.

```
And(?item # ff:Item
    ?item[ff:perishable -> "yes"
          ff:delivered -> ?deliverydate
          ff:scheduled -> ?scheduledate
          ff:deliveredTo -> "John"]
    External(pred:numeric-greater-than(
      External(func:days-from-duration(
        External(func:subtract-dateTimes(
          ?deliverydate ?scheduledate))))

10)))
```

John does not need to know about RIF, however. John can use the OBR rule editor to author the rule and to export it to RIF-BLD XML format. John sends the proposed rule in standard RIF XML format to both of his suppliers.

Jane imports John's rule into an ILOG system, and views the rule using the familiar ILOG syntax. Jane worries about the cost of implementing this rule. She runs a "what-if" analysis using this proposed rule and a sampling of actual delivery data to calculate the expected rate of rejected deliveries.

Jose's Prova inference service receives John's rule, and after querying some past data, decides to propose an alternative rule:

```
if an item is perishable and
   it is delivered to John between 7 and 14 days
   after the scheduled delivery date

then the item will be discounted 18.7%
```

Jose sends the alternate proposal to Jane for comment before replying to John. Jane notices a gap in rule coverage for the case where a delivery is 14 or more days late. Jane uses the ILOG rule editor to add a second rule to the ruleset to cover this case, and sends this final ruleset to John. John imports the ruleset and deploys it to his running supply chain management application with a few mouse clicks.

2 Demo setup

Each participant will have their rule system and RIF translators installed on a laptop computer, as indicated in Fig 1. The RIF XML file will be passed from computer to computer as outlined above. Each participant will show the rules in their native form, and will show the XML content of the RIF XML file. Each participant will also briefly show the rules in action (e.g. part of a BPEL process in Oracle, running a "what-if" scenario in ILOG, or being received, translated, evaluated/executed and refined by a Prova rule inference service deployed on Rule Responder).

Prova

Prova is both a Semantic Web rule language and a highly expressive distributed Semantic Web rule engine which supports, complex reaction rule-based workflows, rule-based complex event processing, distributed Web inference services / agents

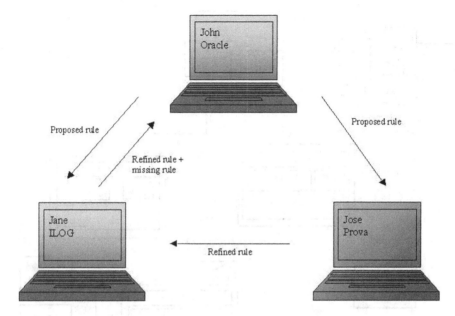

Fig. 1. Rule Interchange Demo Scenario - Interchange using RIF as rule interchange format between three different platform-specific rule execution environments

deployed on an enterprise service bus middleware, rule interchange, declarative (backward reasoning) decision rules and dynamic access to external data sources such as databases, Web services, and Java APIs. Prova follows the spirit and design of the W3C Semantic Web initiative and combines declarative rules, ontologies (vocabularies) and inference with dynamic object-oriented programming and access to external data sources via query languages such as SQL, SPARQL, and XQuery. One of the key advantages of Prova is its elegant separation of logic, data access, and computation as well as its tight integration of Java, Semantic Web technologies and enterprise service-oriented computing and complex event processing technologies.

Figure 2 illustrates the Rule Responder enterprise service middleware [5] that Prova provides for rule interchange. Translator services translate from Prova execution syntax into RIF and vice versa (see Fig. 3). Arbitrary transport protocols such as MS, SMTP, JDBC, TCP, HTTP, and XMPP (the ESB supports all common transport protocols) can be used to transport rule sets, queries and answers between distributed Prova endpoints on the Web using RIF as common rule interchange format.

To specify the semantics of rules, and safeguard the life cycle of rules, including their interchange between different platform-specific rule services, Prova supports test-driven development for self-validating rule bases, where rule sets are interchanged together with test suites (serialized in RIF) which are used for verification and validation (V&V) of the interchanged rules in the target execution environment.

Prova also provides a Rule Manager tools that supports collaborative role-based engineering, project-specific management and runtime execution of Prova projects

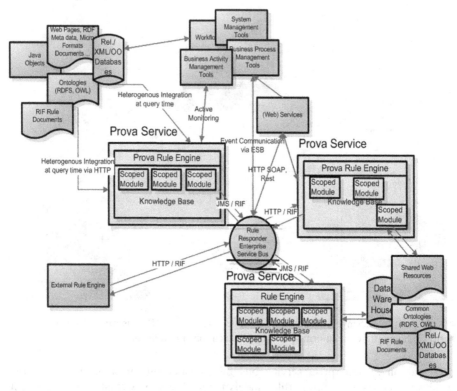

Fig. 2. Prova Rule Engines deployed as Web Rule Inference Services on Rule Responder Enterprise Service Bus using RIF as interchange format

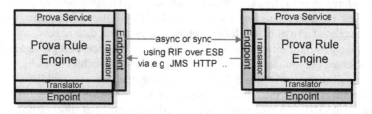

Fig. 3 Prova Rule Engines communicate using RIF via ESB transport (>30 Protocols)

serialized in RIF. The tool splits into two integrated components, the Rule Editor user interface which provides development and testing of rule projects, and the runtime dashboard which enables execution of rules and visualization of results in user-defined (graphical) views.

Prova and RIF

Prova adopts the OMG model driven architecture (MDA) approach:

1. On the computational independent level rules are engineered in the Rule Manager in a natural controlled English language using blueprint templates and user-defined vocabularies.

2. The rules are mapped and serialized in RIF XML which is used as platform independent rule interchange format to interchange rules between Prova inference services and arbitrary other rule execution environments.
3. The RIF rules are translated into the platform specific Prova rule language for execution, which is based on the ISO Prolog syntax and close to the abridged RIF presentation syntax.

The translation between the controlled English rule language and RIF is based on a vocabulary template-driven approach in combination with a controlled English translator. XSLT is used for the translation from RIF into Prova execution syntax (ISO Prolog style).

Online translator web services using the XSLT stylesheet are provided which translate from W3C RIF into Prova. For the translation from Prova back into RIF an Abstract Syntax Tree (AST) using ANTLR v3 is build based on the Prova grammar (http://www.prova.ws/gram.html). The AST is translated into RIF XML syntax. The functionality is also wrapped into online translator services.

The translator services are configured in the transport channels of the inbound and outbound links of the deployed online Prova inference service on the ESB. Incoming RIF rule documents (receive) are translated into Prova rule sets which can be executed by the local Prova rule engine and outgoing rule sets (send) are translated into RIF in the outbound channels before they are transferred via a selected transport protocol such as HTTP or JMS.

ILOG JRules

JRules is the production rule engine and environment developed and marketed by ILOG. It supports the two different levels of a full-fledged Business Rule Management System: the technical level targeted at software developers and the business action language targeted at business users.

JRules supports writing rules in a language close to natural language that can be understood and managed by "business analysts". These Business-level languages are typically compiled into lower-level technical languages; the simple 'production rules' semantics remains unchanged.

The "Business Layer" is composed of additional models supporting the definition of the rules (see Figure 4 below).

- A Business Object Model (BOM) defines the classes representing the objects of the application domain. A BOM is in fact a simplified ontology defining classes with their associated attributes, predicates and actions.
- A Vocabulary model (VOC) defines the words or phrases used to reference the BOM entities.

A Business Action Language (BAL) Rule, close to natural language, is typically an If-Then statement that composes the elements of the VOC to build user understandable rules, such as "**if** the age of a human is greater than 18 **then** the human is major". This rule relates the attributes "age" – a positive integer – and Boolean attribute "major" of the class Human.

The business object model and vocabulary are defined in an early stage of the project, and their projection on an executable layer – the Execution Object Model (XOM) and technical rule language – is implemented by a group of IT specialists. The technical-level layer is used to execute the business rules on the application data within the application architecture.

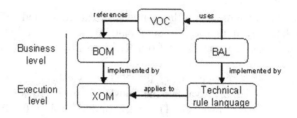

Fig. 4. Different layers of production rules in ILOG

ILOG and RIF

In the demo scenario, we have two complementary objectives: translating a RIF ruleset into the ILOG technical language (IRL) and translating IRL rules into BAL rules. We first produce IRL files from the RIF, and then we produce the BAL-level file from the IRL files. Having rules in distinct levels of abstractions increases the difficulty of round-tripping across the different BRMS.

Rather than having a pure syntactic approach, we base our approach on Model Driven Engineering techniques [7]. We chose to parse the textual source files and to create explicit models. These models are engineered and transformed (using model transformations [8]) into target models that are eventually stored as XML (or text) files conforming to the target format.

The architecture follows the usual *horseshoe* pattern: **inject, transform, extract,** and relies on three core MDE practices and technologies: Domain Specific Languages (DSLs) [7], metamodeling [7], model transformations [8] and projections across technical spaces (i.e., XML, grammar and modeling). DSLs and metamodeling are used to design the rule languages; model transformations are used to produce a set of output models from a set of input models; injections allows translating the input rules (and related artifacts) into models (e.g., text(XML)-to-model translation); and extractions allows translating models into the output artifacts (e.g., model-to-text(XML) translation).

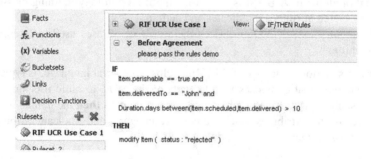

Fig. 5. Oracle Business Rule Designer

We rely on the tools developed at the Atlas Inria project (ATL [8], AM3 [9], and TCS [9]). These tools are developed as Eclipse plug-ins, which enables an easy integration with JRules.

Oracle Business Rules

Oracle Business Rules are part of Oracle Fusion Middleware [10]. OBR includes a production rule engine based on Jess [11], a graphical rule editor (figure 5), and integration with Java applications, service-oriented applications, and Oracle enterprise applications. Integration includes:

- ability to package and deploy rules with applications, and share rules across applications,
- ability to change rules before and after deployment in order to customize applications, and
- support for a wide variety of structured data including Javabeans, XML, and Oracle database-backed business objects.

OBR and RIF

Oracle Business Rules are stored in a proprietary XML format called a Rule Dictionary. We have implemented prototype translators to translate a subset of OBR to/from a subset of RIF Basic Logic Dialect. The subset is a kind of "core" language that should be accepted by many rule systems. **Table 1** shows the correspondence between BLD and OBR.

Table 1. Correspondence between BLD and OBR

BLD feature	BLD example	OBR feature	OBR example
Rule	B :- A	Rule	IF A THEN B
Frame classification and slot test	And(?item#Item ?item[deliveredTo->"John"])	Simple rule test	Item.deliveredTo=="John"
Built in	External(func:numeric-add(?v 1))	Built in	v + 1
Frame formula conclusion	?item[status->"rejected"]	Modify action	Modify Item(status: "rejected")

Following are some restrictions of our prototype when translating from BLD to OBR:

- The "then" part of a rule may contain only a BLD frame formula.
- Only the built in functions and predicates are supported. Logical functions (defined by rules) are not supported.
- Full equality is not supported.
- OBR fact types are really Javabeans. These differ from BLD frames. E.g. in OBR each Item has at most one status. This difference could be expressed as rules, but our prototype does not fully understand such "frame axioms".
- OBR does not support relations, only "frames".

When translating from OBR to BLD, a number of OBR features are not supported because BLD has no equivalent feature:

- Negation is not supported in BLD because of widespread differences in semantics.
- Aggregation is not supported in BLD because it requires a "closed world" semantics.
- Priority and mutual exclusion are not supported in BLD because these are operationally defined.
- User-defined external functions are not supported in BLD because this would require inventing or endorsing a non-logic-based programming language.
- Retraction is not supported because BLD is monotonic.

Destructive modification is not supported, because this is equivalent to retraction. A rule (or chain of rules) may not modify the same properties that it tests.

3 Related Work

There have been several important general standardization or standards-proposing efforts for Rule markup languages acting as lingua franca for exchanging rules between different systems and tools, such as RuleML (www.ruleml.org), SWRL (www.w3.org/Submission/SWRL/), R2ML (oxygen.informatik.tu-cottbus.de/rewerse-i1/?q=R2ML), and others more domain-specific such as RBSLA (Rule Based Service Level Agreement Language (http://ibis.in.tum.de/projects/rbsla/index.php) or SWSL (http://www.w3.org/Submission/SWSF-SWSL/).

RuleML

The Rule Markup Language (RuleML, www.ruleml.org) is a markup language developed to express various families of Web rules in XML for deduction, rewriting, and reaction, as well as further inferential, transformational, and behavioral tasks. It is defined by the Rule Markup Initiative (www.ruleml.org), an open network of individuals and groups from both industry and academia that was formed to develop a canonical Web language for rules using XML markup and transformations from and to other rule standards/systems. It develops a modular, hierarchical specification for different types of rules comprising facts, queries, derivation rules, integrity constraints (consistency-maintenance rules), production rules, and reaction rules (Reaction RuleML, http://ibis.in.tum.de/research/ReactionRuleML), as well as tools and transformations from and to other rule standards/systems. The RuleML Initiative is also heavily involved in the development of W3C RIF.

SWRL

The Semantic Web Rule Language (SWRL, www.w3.org/Submission/SWRL/) is a proposal for a Semantic Web rule language combining sublanguages of the OWL Web Ontology Language (OWL DL and Lite) with those of the Rule Markup Language (Unary/Binary Datalog). The specification was submitted in May 2004 to the W3C by the National Research Council of Canada, Network Inference (since acquired by webMethods), and Stanford University in association with the Joint US/EU ad hoc Agent Markup Language Committee.

R2ML

R2ML (http://oxygen.informatik.tu-cottbus.de/rewerse-i1/?q=R2ML) was developed as a subproject in the EU Network of Excellence REWERSE (http:// oxygen. informatik.tu-cottbus.de/rewerse-i1/). The R2ML project is about the design of integrity and derivation rules on the basis of the Rule Markup Language (RuleML) and the Semantic Web Rule Language (SWRL).

4 Conclusion

We have demonstrated that W3C RIF BLD provides a robust core for the interchange of business rules among commercial and academic rule systems embodying both production rules and derivation rules.

We will continue to work to create additional dialects, such as the Production Rule Dialect [13] and Reaction Rules Dialect [14], to increase the scope of RIF and support more features of ILOG, OBR, Prova, and other rule systems.

References

1. W3C Rule Interchange Format,
   http://www.w3.org/2005/rules/wiki/RIF_Working_Group
2. ILOG JRules. Ref. site (April 2008),
   http://www.ilog.com/products/jrules/index.cfm
3. Oracle Business Rules,
   http://www.oracle.com/technology/products/ias/
   business_rules/index.html
4. Prova, http://www.prova.ws
5. Rule Responder, http://responder.ruleml.org/
6. RIF Basic Logic Dialect, http://www.w3.org/TR/rif-bld/
7. Kurtev, I., Bézivin, J., Jouault, F., Valduriez, P.: Model-based DSL Frameworks. In: Proc of. Companion of OOPSLA 2006, Portland, OR, USA, October 22-26, 2006, pp. 602–616 (2006)
8. Jouault, F., Kurtev, I.: Transforming Models with ATL. In: Proc. of the Model Transformations in Practice Workshop at MoDELS 2005 (2005)
9. Eclipse Generative Modeling Technologies (GMT),
   http://www.eclipse.org/gmt/
10. Oracle Fusion Middleware,
    http://www.oracle.com/technology/products/middleware/
    index.html
11. Jess, http://herzberg.ca.sandia.gov/jess/
12. W3C RIF Use Cases and Requirements, http://www.w3.org/TR/rif-ucr/
13. W3C RIF Production Rule Dialect, http://www.w3.org/TR/rif-prd/
14. W3C RIF Reaction Rule Dialect, http://www.w3.org/2005/rules/wiki/RRD

Self-sustained Routing for Event Diffusion in Wireless Sensor Networks

Kirsten Terfloth and Jochen Schiller

Institute of Mathematics and Computer Science,
Takustr. 9, 14195 Berlin, Germany
{terfloth,schiller}@inf.fu-berlin.de
http://cst.mi.fu-berlin.de

Abstract. Wireless sensor networks have the potential to become a scalable, low-cost and highly flexible tool for distributed monitoring and/or event recognition: Embedded devices that coordinate themselves via wireless communication to implement a common, distributed application are already seeing first industrial adaption in e.g. home automation, personal health or environmental monitoring. Despite of their valuable properties they put a high burden upon application development since critical issues such as a general resource scarcity, the unreliable communication medium and the management of distribution are often visible throughout the protocol stack to ensure efficient utilization.

In the demo, we will show how our middleware framework FACTS helps to alleviate problems in implementation of both system- and application-level code. Providing a rule-based, domain-specific language, FACTS is especially suited to express event-driven tasks. Therefore, we will showcase how to efficiently program a self-sustained routing protocol forwarding relevant events to dedicated sink nodes with only a handful of rules.

Keywords: domain-specific language, routing, wireless sensor networks, middleware, rules, FACTS.

1   Introduction

Wireless sensor networks have become an increasingly important tool to monitor environmental or industrial processes. To set up a working system, numerous sensor nodes are deployed in the field to cooperatively perform a specific task, such as reporting of measurements or uncommon events. These embedded devices usually feature a variety of sensors to measure physical parameters of their surrounding, a micro controller to process the data obtained and a transceiver allowing for communication with other nodes in their vicinity to organize themselves into a wireless ad-hoc network. Since the lifetime of such a network powered by batteries is bound by the energy spend per node, an event-centric approach is very popular in practice in order to prolong network duty.

In summary, software development for WSNs is inherently difficult - the highly embedded nature of the nodes, the event-centric way of processing data and the collaborative, distributed manner of deployments call for expertise in all three

N. Bassiliades, G. Governatori, and A. Paschke (Eds.): RuleML 2008, LNCS 5321, pp. 236–241, 2008.
© Springer-Verlag Berlin Heidelberg 2008

domains to be able to obtain a fully-functional network. To overcome these challenges arising from the necessity to take a very system-oriented view on application development, FACTS, a rule-based middleware framework has been developed. Capturing event-centricity at the language level, a programmer is able to express reactions to new sensor readings, changes in the state of a sensor node or incoming messages with the help of rules that are evaluated on the sensor nodes.

Since a viable routing infrastructure is mandatory to enable both information delivery to dedicated sink nodes in the network as well as to maintain group connectivity for distributed data evaluation, we picked a use case implementation from this domain. Our demo will present a light-weight implementation of the Directed Diffusion paradigm [1], a popular data-centric information diffusion technique for wireless sensor networks. We enhanced the protocol in a way so that it supports streaming of event data from multiple sources to multiple sinks, and show, how simple it is to map routing decisions to rules.

The rest of this paper is structured as follows: In the next section, we sketch the target domain for our framework and review hardware prerequisites that influenced design decisions. Before turning to the actual core of the demo in Section 3, we also briefly introduce language and framework features to establish a common ground, before concluding the paper in Section 4.

2   Ruling Sensor Networks with FACTS

Unlike web-based application areas for rule systems, wireless sensor nodes primarily suffer from memory shortage. The ScatterWeb sensor nodes as depicted in Figure 1 that serve as our experimental basis i.e. offers 5 KB of RAM and 55KB of program flash to operate, thus demand for efficient usage of these resources. Naturally, instead of offering a full-fledged, general-purpose but resource-intense operating system, the firmware is optimized to only implement the absolute minimum of necessary functionality. This way, essential drivers for the sensors (accelerometer, humidity and temperature), the Chipcon CC1020 transceiver including a CSMA MAC protocol as well as support for using the serial interface for debugging and command prompting are available.

2.1   FACTS Architecture

The motivation for FACTS has been to offer an abstraction from having to deal with programming embedded devices as described above, thus shift the burden of e.g. memory management and event scheduling from the programmer's responsibility to a runtime environment. The usage of rules as a programming entity has two advantages: First of all, specification of actions corresponding to a dedicated event and current state of the node maps very naturally to the foremost operational model of duty cycling in WSNs: For reasons of energy efficiency, nodes should only respond to and process relevant changes in their environment, and otherwise switch to a low-power mode. Furthermore, the modularity of rules

Fig. 1. The MSB430 sensor node

can be used to allow for energy-efficient software updates of a running system: Replacement of a new routing protocol may be done by substitution of a couple of rules instead of reflashing a complete image onto a device.

The middleware idea makes use of exactly this: the vision is to bundle sets of interacting rules into a ruleset and supply it to a user as a service. Some applications may need a routing service, others a reliable neighborhood abstraction, etc. - each of which is wrapped into a ruleset and may be used by application programmers as necessary. Evaluation of rules using the rule engine will be triggered whenever a fact in the repository, which is the storage entity for facts, is altered or added. This can be the case either after a new sensor reading is wrapped as a fact by the firmware, a fact has been received over the radio interface of a sensor node or it has been produced as an output of a previous rule execution. Rules have to be given a priority to specify the order of execution, but allow for multiple or hierarchical reactions to one event. Rule execution thus stops as soon as no change has occurred in the fact repository in the last round of rule checking or no rule is associated to the last change. The runtime environment will then be simply set to sleep.

2.2   RDL - The Core of Facts

The idea of describing node behavior with a couple of rules stating how to react to new data items has to be expressible. Therefore, a new language that features dedicated instructions has been proposed, offering four basic constructs:

- **Rules.** A rule states under which circumstances `condition` what kind of action or `statement` will be triggered.
- **Facts.** A fact is the representation of data, both on the system and within the network. Basically, it is a named data tuple with both inherent properties automatically created at creation time such as a timestamps, a flag whether

it has been modified, etc., and possibly explicitly defined properties by the programmer.

- **Slots.** A slot is a named filter that acts upon the fact repository, thus is evaluated at runtime. Multiple facts may match a slot, thus statements applied to slots will be applied to all matching facts.
- **Functions.** A function provides a hook into the firmware of a sensor node, thus allows a programmer to specify system-inherent source code not feasible to express in a rule-based manner in plain C instead. This may include for instance printing on the serial and can be called within a statement.

For more details concerning the expressiveness of the language see [2].

3 Self-sustained Routing with FACTS

In our demo, we will set up a testbed of ten ScatterWeb MSB430 wireless sensor nodes, each of which tasked via rules to participate in the routing process to forward events recognized by a dedicated source node. Informally, the idea of the directed diffusion routing paradigm is to enable streaming of information from one or more data sources to a sink node that subscribed for a certain type of information in absence of unique network ids of sensor nodes. The basic variant, commonly referred to as *one phase pull* starts with interest dissemination followed by simple information diffusion. First, a sink announces its interest, thus a message containing data type and rate to be delivered back, by flooding it into the network. Upon reception of an interest, a node establishes a so called *gradient* in its gradient cache pointing towards the sending node to establish a backward path. Reinforcement strategies can be applied to ensure choosing the shortest path to the sink. A potential data source receiving a request will start reporting available data, thus route it back to the inquirer using only its local information. Since the wireless medium implies unstable links between individual nodes, periodic maintenance prevents network failure.

Implementing directed diffusion using RDL is a straight-forward task: each decision that a node has to make upon the reception of a packet can be easily mapped to a single rule, dependent on received fact and current state of the node. Listing 1.1 shows the four rules necessary to handle interests coming in from another node: Drop the interest in case the node is itself a sink (lines 6-9), update the routing information stored in the gradient in case the hops the received interest traveled is smaller than the stored value and resend the interest (lines 11-23), drop the interest in case similar information has been seen before (lines 25-30) and finally setup a new gradient in case the node received this type of interest for the first time. The complete ruleset which comprises interest dissemination, gradient setup, route maintenance and data propagation to multiple sinks can be expressed in 14 rules, solely occupying 1682 Byte of flash. Assuming another couple of rules in addition for application

Listing 1.1. Processing interests with the help of rules in FACTS

```
1
2   /*
3    * fact interest: interest [sink (int), weight (int),
          neighbor (int)]
4    */
5
6   rule dropInterestsWhenSink 96
7   <- exists {interest
8       <- eval ({this sink} == systemID)}
9   -> retract {interest}
10
11  rule reinforceInterests 95
12  <- exists {interest}
13  <- exists {gradient
14      <- eval ({this sink} == {interest sink})
15      <- eval ({this weight} > {interest weight})
16  }
17  -> set {gradient weight
18      <- eval ({this sink} == {interest sink})
19  } = {interest weight}
20  -> set {interest neighbor} = systemID
21  -> set {interest weight} = ({interest weight} + 1)
22  -> send systemBroadcast systemTxRange {interest}
23  -> retract {interest}
24
25  rule removeDuplicateInterests 94
26  <- exists {interest}
27  <- exists {gradient
28      <- eval ({this sink} == {interest sink})
29  }
30  -> retract {interest}
31
32  rule handleInterests 93
33  <- exists {interest}
34  -> define gradient [sink = {interest sink}, neighbor = {
          interest neighbor}, weight = {interest weight}]
35  -> set {interest neighbor} = systemID
36  -> set {interest weight} = ({interest weight} + 1)
37  -> send systemBroadcast systemTxRange {interest}
38  -> retract {interest}
```

logic such as event recognition and filtering, this example shows that rules provide a very concise expressiveness especially valuable in the embedded domain.

Our demo will provide a step-by-step overview on how to set up a running sensor network testbed as well as show the actual event diffusion to two distinct sink nodes.

4 Conclusions

Running a rule engine on embedded devices can provide a great deal of abstraction to non-experts due to the clear separation of concerns when using rules. With our demo showing how to provide a self-sustained routing algorithm in just a couple of rules, we'd like to point out an interesting new field of application for forward chaining mechanisms and rule adaption that yet has received little attention.

References

1. Intanagonwiwat, C., Govindan, R., Estrin, D.: Directed diffusion: A scalable and robust communication paradigm for sensor networks. In: Mobile Computing and Networking, pp. 56–67 (2000)
2. Terfloth, K., Wittenburg, G., Schiller, J.: Rule-oriented programming for wireless sensor networks. In: Proceedings of the International Conference on Distributed Computing in Sensor Networks (DCOSS)/ EAWMS Workshop (June 2006)

Author Index